Math Magicians: Machine Learning Calculus Skills Practice Workbook

Jamie Flux

https://www.linkedin.com/company/golden-dawn-engineering/

Collaborate with Us!

Have an innovative business idea or a project you'd like to collaborate on?
We're always eager to explore new opportunities for growth and partnership.
Please feel free to reach out to us at:

https://www.linkedin.com/company/golden-dawn-engineering/

We look forward to hearing from you!

Contents

Chapter 1

Functions and Graphs

Practice Problems 1

1. Consider the function $f : \mathbb{R} \to \mathbb{R}$ defined by $f(x) = x^3 - 6x + 4$. Determine the coordinates of the points where the function intersects the x-axis.

2. For the function $g(x) = \frac{x^2-9}{x-3}$, determine whether it is continuous at $x = 3$. Explain your reasoning.

3. If $h(x) = x^2 + 4x$, evaluate $h(x)$ at the critical points to determine the local minima and maxima.

4. Determine whether the function $f(x) = \frac{1}{x}$ has any asymptotes, and identify their locations.

5. Analyze the symmetry of the function $f(x) = x^5 - 4x^3$ to determine if it is odd, even, or neither.

6. Consider the exponential function $f(x) = 3e^{2x}$. Find the y-intercept and the horizontal asymptote of the graph of this function.

Answers 1

1. **Solution:** To find the x-intercepts of $f(x) = x^3 - 6x + 4$, we set $f(x) = 0$:

$$x^3 - 6x + 4 = 0$$

This cubic equation can be solved using the Rational Root Theorem or numerical method approximation, identifying roots (x-intercepts) at $x = 2$, $x = -2$ and $x = 1.333$. The exact roots would be obtained using factoring or approximation techniques.

2. **Solution:** For $g(x) = \frac{x^2 - 9}{x - 3}$, substitute $x = 3$:

$$g(x) = \frac{(x-3)(x+3)}{x-3} = x + 3, \quad x \neq 3$$

Therefore, $g(3)$ is not defined directly, but the limit as $x \to 3$ is:

$$\lim_{x \to 3} g(x) = 3 + 3 = 6$$

Thus, we define $\tilde{g}(x) = x + 3$ for continuity at $x = 3$.

3. **Solution:** Critical points occur where the derivative $h'(x) = 2x + 4$ is zero:

$$2x + 4 = 0 \quad \Rightarrow \quad x = -2$$

Evaluate $h(x)$ at $x = -2$:
$$h(-2) = (-2)^2 + 4(-2) = 4 - 8 = -4$$

This point is a local minimum.

4. **Solution:** The function $f(x) = \frac{1}{x}$:
 - Vertical asymptote at $x = 0$, since $f(x)$ is not defined.
 - No horizontal asymptote as $\lim_{x \to \infty} f(x) = 0$.

5. **Solution:** Check symmetry for $f(x) = x^5 - 4x^3$:
   - Even: $f(-x) = (-x)^5 - 4(-x)^3 = -x^5 + 4x^3 \neq f(x)$
   - Odd: $f(-x) = -f(x)$ holds, so the function is odd.

6. **Solution:** For $f(x) = 3e^{2x}$:
   - Y-intercept: Set $x = 0$ to find $f(0) = 3e^0 = 3$.
   - Horizontal asymptote does not exist, as e^{2x} grows without bound. However, sometimes approached as $x \to -\infty$, $f(x) \to 0$ can resemble an asymptotic behavior near the x-axis.

Practice Problems 2

1. Determine whether the following function is injective, surjective, or bijective:

$$f(x) = 3x + 2$$

2. Identify the domain and range of the function:

$$g(x) = \sqrt{x - 4}$$

3. Describe the asymptotic behavior of the function:

$$h(x) = \frac{1}{x - 2}$$

9

4. Determine the symmetry of the function:
$$k(x) = x^4 - x^2$$

5. Find the intercepts of the function:
$$m(x) = x^2 - 6x + 8$$

6. Given the function is a polynomial, provide its degree and possible shape:
$$n(x) = 5x^3 - 3x + 1$$

Answers 2

1. Determine whether the following function is injective, surjective, or bijective:
$$f(x) = 3x + 2$$

Solution: A function is injective if distinct inputs produce distinct outputs. For $f(x) = 3x+2$, suppose $f(a) = f(b)$, then $3a + 2 = 3b + 2$, leading to $a = b$. Hence, it is injective.

A function is surjective if every value in the codomain $\mathbb{R}$ can be obtained. For $f(x) = 3x + 2$, for any $y \in \mathbb{R}$, solving $3x + 2 = y$, gives $x = \frac{y-2}{3}$, thus it is surjective.

Since $f(x)$ is both injective and surjective, it is bijective.

2. Identify the domain and range of the function:

$$g(x) = \sqrt{x - 4}$$

Solution: The domain of $g(x) = \sqrt{x - 4}$ requires the expression under the square root to be non-negative:

$$x - 4 \geq 0 \implies x \geq 4$$

Therefore, the domain is $[4, \infty)$.

The range of a square root function is non-negative values, therefore the range is $[0, \infty)$.

3. Describe the asymptotic behavior of the function:

$$h(x) = \frac{1}{x - 2}$$

Solution: The vertical asymptote occurs where the function is undefined, which is at $x = 2$.

As $x \to 2^+$, $h(x) \to \infty$, and as $x \to 2^-$, $h(x) \to -\infty$.

The horizontal asymptote is determined by the behavior as $x \to \pm\infty$. Here, $h(x) \to 0$.

4. Determine the symmetry of the function:

$$k(x) = x^4 - x^2$$

Solution: Evaluating symmetry:
- Even symmetry: $k(-x) = (-x)^4 - (-x)^2 = x^4 - x^2 = k(x)$, hence the function is even.
- Thus, there is symmetry about the y-axis.

5. Find the intercepts of the function:

$$m(x) = x^2 - 6x + 8$$

Solution: To find x-intercepts, set $m(x) = 0$:

$$x^2 - 6x + 8 = 0$$

Using the quadratic formula:

$$x = \frac{-b \pm \sqrt{b^2 - 4ac}}{2a} = \frac{6 \pm \sqrt{36 - 32}}{2} = \frac{6 \pm 2}{2}$$

Solutions are $x = 4$ and $x = 2$.

The y-intercept, setting $x = 0$:

$$m(0) = 0^2 - 6 \cdot 0 + 8 = 8$$

Thus, intercepts are $(4, 0)$, $(2, 0)$, and $(0, 8)$.

6. Given the function is a polynomial, provide its degree and possible shape:

$$n(x) = 5x^3 - 3x + 1$$

Solution: The degree of the polynomial is 3, which indicates it is a cubic function.

In general, cubic polynomials can have one real root or three real roots and often have inflection points where the direction of curvature changes. Being degree 3, they lack symmetry.

Practice Problems 3

1. Consider the function $f(x) = 3x^3 - 2x + 1$. Determine whether it is injective, surjective, or bijective from $\mathbb{R} \to \mathbb{R}$.

2. Analyze the symmetry of the function $f(x) = x^4 - x^2$.

3. Determine the intercepts of the function $f(x) = x^2 - 4x + 3$.

4. For the function $f(x) = \frac{x-1}{x+2}$, discuss the asymptotic behavior.

5. Determine the domain and range of the function $f(x) = \sqrt{x-1} + 2$.

6. Given the function $f(x) = e^x + \ln(x)$, find the domain of f when defined from $\mathbb{R}^+ \to \mathbb{R}$.

Answers 3

1. Consider the function $f(x) = 3x^3 - 2x + 1$. Determine whether it is injective, surjective, or bijective from $\mathbb{R} \to \mathbb{R}$.

   Solution:

   To determine injectivity, check if the derivative $f'(x) = 9x^2 - 2$ is always positive or always negative since f is continuous and differentiable. Here:

   $$f'(x) = 9x^2 - 2$$

   For injectivity, $f'(x) \neq 0$ for all x:

   $$9x^2 - 2 = 0 \implies x^2 = \frac{2}{9} \implies x = \pm\frac{\sqrt{2}}{3}$$

   Since $f'(x)$ changes sign at these points, f is not injective.

   To check surjectivity, note that $\lim_{x \to \pm\infty} f(x) = \pm\infty$. Thus, $f(x)$ is surjective because it covers all reals.

   Since $f(x)$ is surjective but not injective, it is not bijective.

2. Analyze the symmetry of the function $f(x) = x^4 - x^2$.

   Solution:

   To determine symmetry:

   Check if $f(-x) = f(x)$ (even function):

   $$f(-x) = (-x)^4 - (-x)^2 = x^4 - x^2 = f(x)$$

   Thus, $f(x)$ is even.

3. Determine the intercepts of the function $f(x) = x^2 - 4x + 3$.

   Solution:

   x-intercepts are found by setting $f(x) = 0$:

   $$x^2 - 4x + 3 = 0 \implies (x - 1)(x - 3) = 0$$

   Thus, $x = 1$ and $x = 3$.

   The y-intercept is found by evaluating $f(0) = 0^2 - 4 \times 0 + 3 = 3$.

   Therefore, intercepts are $(1, 0), (3, 0)$ for x-intercepts, and $(0, 3)$ for the y-intercept.

4. For the function $f(x) = \frac{x-1}{x+2}$, discuss the asymptotic behavior.

   Solution:

   Vertical asymptotes occur where the denominator is zero:

   $$x + 2 = 0 \implies x = -2$$

   Horizontal asymptotes are determined by the leading coefficients as $x \to \pm\infty$:

   $$\lim_{x \to \pm\infty} \frac{x-1}{x+2} = \lim_{x \to \pm\infty} \frac{1 - \frac{1}{x}}{1 + \frac{2}{x}} = 1$$

   Therefore, $y = 1$ is a horizontal asymptote.

5. Determine the domain and range of the function $f(x) = \sqrt{x-1} + 2$.

   Solution:

   The domain of $f(x) = \sqrt{x-1} + 2$ is when the expression inside the square root is non-negative:

   $$x - 1 \geq 0 \implies x \geq 1$$

   So, the domain is $[1, \infty)$.

   For the range:

   $$f(x) = \sqrt{x-1} + 2 \quad \text{where} \quad x \geq 1$$
   $$\Rightarrow f(1) = 0 + 2 = 2 \quad \text{and as} \quad x \to \infty, f(x) \to \infty.$$

   Thus, the range is $[2, \infty)$.

6. Given the function $f(x) = e^x + \ln(x)$, find the domain of f when defined from $\mathbb{R}^+ \to \mathbb{R}$.

   Solution:

   The function e^x is defined for all $x \in \mathbb{R}$, whereas $\ln(x)$ is defined for $x > 0$.

   Therefore, the domain of $f(x) = e^x + \ln(x)$ is $(0, \infty)$.

Chapter 2

Limits and Continuity

Practice Problems 1

1. Evaluate the following limit using the epsilon-delta definition:

$$\lim_{x \to 2}(3x - 4) = 2$$

2. Determine the limit as x approaches infinity:

$$\lim_{x \to \infty} \frac{5x^2 + 3x}{2x^2 - x}$$

3. Calculate the limit using L'Hôpital's Rule:

$$\lim_{x \to 0} \frac{\sin(3x)}{\tan(2x)}$$

4. Use continuity principles to determine the existence of the following:

$$\lim_{x \to 1} \frac{x^3 - 1}{x - 1}$$

5. Analyze the type of discontinuity present at $x = 2$ for the function:

$$f(x) = \frac{x^2 - 4}{x - 2}$$

6. Verify the Intermediate Value Theorem for the function $f(x) = x^3 - 3x + 1$ on the interval $[-2, 0]$ to show there is a root within the interval.

Answers 1

1. **Solution:** We start with the epsilon-delta definition for limits. Given ϵ, we want:

$$|f(x) - L| < \epsilon \implies |3x - 4 - 2| < \epsilon$$

Simplifying, we need:

$$|3x - 6| < \epsilon \implies 3|x - 2| < \epsilon \implies |x - 2| < \frac{\epsilon}{3}$$

Choose $\delta = \frac{\epsilon}{3}$. Thus for $|x - 2| < \delta$, the limit exists with $|(3x - 4) - 2| < \epsilon$.

2. **Solution:** To find the limit $\lim_{x \to \infty} \frac{5x^2 + 3x}{2x^2 - x}$:

$$= \frac{5 + \frac{3}{x}}{2 - \frac{1}{x}}$$

As $x \to \infty$, $\frac{3}{x} \to 0$ and $\frac{1}{x} \to 0$. Thus:

$$\lim_{x \to \infty} \frac{5x^2 + 3x}{2x^2 - x} = \frac{5}{2}$$

3. **Solution:** L'Hôpital's Rule applies as $\frac{0}{0}$ form. Differentiate numerator and denominator:

$$\lim_{x \to 0} \frac{3\cos(3x)}{2\sec^2(2x)}$$

Evaluating at $x = 0$ gives:

$$= \frac{3\cos(0)}{2 \cdot 1} = \frac{3}{2}$$

4. **Solution:** Applying direct substitution:

$$\lim_{x \to 1} \frac{x^3 - 1}{x - 1} = \lim_{x \to 1} \frac{(x-1)(x^2 + x + 1)}{x - 1}$$

Cancel the $x - 1$:

$$= \lim_{x \to 1} (x^2 + x + 1) = 1 + 1 + 1 = 3$$

5. **Solution:** The function is $\frac{x^2 - 4}{x - 2} = \frac{(x-2)(x+2)}{x-2}$.

$$\lim_{x \to 2} (x + 2) = 4$$

This is a removable discontinuity since redefining $f(2) = 4$ restores continuity.

6. **Solution:** $f(x) = x^3 - 3x + 1$ is continuous. Check values:

$$f(-2) = -8 + 6 + 1 = -1, \quad f(0) = 1$$

As $-1 \leq 0 \leq 1$, by IVT, a root exists in $[-2, 0]$.

Practice Problems 2

1. Evaluate the following limit:

$$\lim_{x \to 2} \frac{x^2 - 4}{x - 2}$$

2. Determine the limit at infinity:

$$\lim_{x \to \infty} \frac{3x^2 + 5}{2x^2 - x}$$

3. Verify the continuity of the function at $x = 1$:

$$f(x) = \begin{cases} x^2 + 3, & \text{if } x < 1 \\ 4x - 1, & \text{if } x \geq 1 \end{cases}$$

4. Use L'Hôpital's Rule to find the following limit:

$$\lim_{x \to 0} \frac{\sin(x)}{x}$$

5. Identify the type of discontinuity at $x = 2$:

$$g(x) = \frac{x^2 - 4}{x - 2}$$

6. Check if the Intermediate Value Theorem applies to the function $h(x) = x^3 - 5x + 3$ on the interval $[-1, 2]$ and determine c such that $h(c) = 0$.

Answers 2

1. Evaluate the following limit:
$$\lim_{x \to 2} \frac{x^2 - 4}{x - 2}$$

   Solution: Factor the numerator:
   $$x^2 - 4 = (x - 2)(x + 2)$$

   The expression becomes:
   $$\frac{(x - 2)(x + 2)}{x - 2}$$

   Cancel $(x - 2)$, provided $x \neq 2$:
   $$= x + 2$$

   Thus,
   $$\lim_{x \to 2} (x + 2) = 4$$

2. Determine the limit at infinity:
$$\lim_{x \to \infty} \frac{3x^2 + 5}{2x^2 - x}$$

   Solution: Divide the numerator and denominator by x^2:
   $$= \lim_{x \to \infty} \frac{3 + \frac{5}{x^2}}{2 - \frac{1}{x}}$$

   As $x \to \infty$, $\frac{5}{x^2} \to 0$ and $\frac{1}{x} \to 0$:
   $$= \frac{3 + 0}{2 - 0} = \frac{3}{2}$$

3. Verify the continuity of the function at $x = 1$:
$$f(x) = \begin{cases} x^2 + 3, & \text{if } x < 1 \\ 4x - 1, & \text{if } x \geq 1 \end{cases}$$

   Solution: Check $\lim_{x \to 1^-} f(x)$:
   $$\lim_{x \to 1^-} (x^2 + 3) = 1^2 + 3 = 4$$

   Check $\lim_{x \to 1^+} f(x)$:
   $$\lim_{x \to 1^+} (4x - 1) = 4(1) - 1 = 3$$

   $f(1) = 4(1) - 1 = 3$. Since the limits differ, Function is not continuous at $x = 1$.

4. Use L'Hôpital's Rule to find the following limit:

$$\lim_{x \to 0} \frac{\sin(x)}{x}$$

Solution: The form $\frac{0}{0}$ is indeterminate: use L'Hôpital's Rule:

$$= \lim_{x \to 0} \frac{\cos(x)}{1} = \cos(0) = 1$$

5. Identify the type of discontinuity at $x = 2$:

$$g(x) = \frac{x^2 - 4}{x - 2}$$

Solution: Factor $x^2 - 4 = (x - 2)(x + 2)$:

$$g(x) = x + 2 \text{ when } x \neq 2$$

Removable discontinuity at $x = 2$ since limit exists:

$$\lim_{x \to 2} g(x) = 4$$

6. Check if the Intermediate Value Theorem applies to the function $h(x) = x^3 - 5x + 3$ on the interval $[-1, 2]$ and determine c such that $h(c) = 0$.
   Solution: Calculate $h(-1)$ and $h(2)$:

$$h(-1) = (-1)^3 - 5(-1) + 3 = -1 + 5 + 3 = 7$$

$$h(2) = (2)^3 - 5(2) + 3 = 8 - 10 + 3 = 1$$

$h(x)$ transitions from 7 to 1 as x moves from -1 to 2, hence passes zero (Intermediate Value Theorem). By continuity, there exists $c \in [-1, 2]$ such that $h(c) = 0$.

Practice Problems 3

1. Evaluate the following limit using the epsilon-delta definition:

$$\lim_{x \to 3} (2x + 1)$$

2. Determine whether the function is continuous at $x = 2$:

$$f(x) = \begin{cases} x^2 & \text{if } x \neq 2 \\ 5 & \text{if } x = 2 \end{cases}$$

3. Evaluate the limit using factorization:
$$\lim_{x \to 2} \frac{x^2 - 4}{x - 2}$$

4. Use L'Hôpital's Rule to find the limit:
$$\lim_{x \to 0} \frac{\sin(x)}{x}$$

5. Identify the type of discontinuity (if any) at $x = 1$ for the function:
$$g(x) = \frac{x^2 - 1}{x - 1}$$

6. Verify the Intermediate Value Theorem for the function between $x = 1$ and $x = 3$:

$$h(x) = x^3 - 3x + 1$$

and $L = 0$.

Answers 3

1. **Solution:** To evaluate $\lim_{x \to 3}(2x + 1)$ using the epsilon-delta definition, we set:

$$L = 2 \cdot 3 + 1 = 7$$

Given $\epsilon > 0$, choose $\delta = \epsilon/2$. If $0 < |x - 3| < \delta$, then:

$$|(2x + 1) - 7| = |2x - 6| = 2|x - 3| < 2\delta = \epsilon$$

Therefore, the epsilon-delta condition is satisfied, confirming that:

$$\lim_{x \to 3}(2x + 1) = 7.$$

2. **Solution:** Check continuity at $x = 2$ for $f(x)$:

$$\lim_{x \to 2} x^2 = 2^2 = 4$$

Since $\lim_{x \to 2} f(x) \neq f(2)$ ($4 \neq 5$), $f(x)$ is not continuous at $x = 2$.

3. **Solution:** Factor the numerator:

$$\lim_{x \to 2} \frac{x^2 - 4}{x - 2} = \lim_{x \to 2} \frac{(x - 2)(x + 2)}{x - 2} = \lim_{x \to 2}(x + 2) = 4$$

4. **Solution:** Using L'Hôpital's Rule:

$$\lim_{x \to 0} \frac{\sin(x)}{x} = \lim_{x \to 0} \frac{\cos(x)}{1} = \cos(0) = 1$$

5. **Solution:** Simplify $g(x)$:

$$g(x) = \frac{(x - 1)(x + 1)}{x - 1} = x + 1 \quad \text{for } x \neq 1$$

$$\lim_{x \to 1} g(x) = \lim_{x \to 1}(x + 1) = 2$$

Redefining $g(1) = 2$ removes the discontinuity, indicating a removable discontinuity at $x = 1$.

6. **Solution:** Check $h(x)$ between $x = 1$ and $x = 3$:

$$h(1) = 1^3 - 3 \cdot 1 + 1 = -1, \quad h(3) = 3^3 - 3 \cdot 3 + 1 = 19$$

Since 0 is between -1 and 19, and $h(x)$ is continuous on $[1, 3]$, by the Intermediate Value Theorem, there exists $c \in [1, 3]$ with $h(c) = 0$.

Chapter 3

Basics of Differentiation

Practice Problems 1

1. Differentiate the following function:

$$f_1(x) = 7x^6 - 5x + 3x^3$$

2. Differentiate the following function:

$$f_2(x) = \ln(x^2) + x^{-3} + 4$$

3. Differentiate the following function:

$$f_3(x) = \frac{x^5}{3} - \cos(x) + e^x$$

4. Differentiate the following function:

$$f_4(x) = \tan(x) + x^4 - \frac{1}{x}$$

5. Differentiate the following function:

$$f_5(x) = \sin^2(x) + \frac{1}{2}x^3$$

6. Differentiate the following function:

$$f_6(x) = \frac{x^2 + 1}{x}$$

Answers 1

1. Differentiate the following function:

$$f_1(x) = 7x^6 - 5x + 3x^3$$

Solution:

$$f_1'(x) = \frac{d}{dx}(7x^6) - \frac{d}{dx}(5x) + \frac{d}{dx}(3x^3)$$

$$= 7 \cdot 6x^{6-1} - 5 \cdot 1x^{1-1} + 3 \cdot 3x^{3-1}$$

$$= 42x^5 - 5 + 9x^2$$

Therefore,

$$f_1'(x) = 42x^5 + 9x^2 - 5.$$

2. Differentiate the following function:

$$f_2(x) = \ln(x^2) + x^{-3} + 4$$

Solution:

$$f_2'(x) = \frac{d}{dx}(\ln(x^2)) + \frac{d}{dx}(x^{-3}) + \frac{d}{dx}(4)$$

Using the chain rule for $\ln(x^2)$:

$$= \frac{1}{x^2} \cdot 2x + (-3)x^{-3-1} + 0$$

$$= \frac{2}{x} - 3x^{-4}$$

Therefore,

$$f_2'(x) = \frac{2}{x} - \frac{3}{x^4}.$$

3. Differentiate the following function:

$$f_3(x) = \frac{x^5}{3} - \cos(x) + e^x$$

Solution:

$$f_3'(x) = \frac{d}{dx}\left(\frac{1}{3}x^5\right) - \frac{d}{dx}(\cos(x)) + \frac{d}{dx}(e^x)$$

$$= \frac{1}{3} \cdot 5x^{5-1} + \sin(x) + e^x$$

$$= \frac{5}{3}x^4 + \sin(x) + e^x$$

Therefore,

$$f_3'(x) = \frac{5}{3}x^4 + \sin(x) + e^x.$$

4. Differentiate the following function:

$$f_4(x) = \tan(x) + x^4 - \frac{1}{x}$$

Solution:

$$f_4'(x) = \frac{d}{dx}(\tan(x)) + \frac{d}{dx}(x^4) - \frac{d}{dx}(x^{-1})$$

$$= \sec^2(x) + 4x^{4-1} + x^{-2}$$

$$= \sec^2(x) + 4x^3 + \frac{1}{x^2}$$

Therefore,

$$f_4'(x) = \sec^2(x) + 4x^3 + \frac{1}{x^2}.$$

5. Differentiate the following function:

$$f_5(x) = \sin^2(x) + \frac{1}{2}x^3$$

Solution: Using the chain rule on $\sin^2(x)$:

$$f_5'(x) = \frac{d}{dx}(\sin^2(x)) + \frac{d}{dx}\left(\frac{1}{2}x^3\right)$$

$$= 2\sin(x) \cdot \cos(x) + \frac{1}{2} \cdot 3x^{3-1}$$

$$= 2\sin(x)\cos(x) + \frac{3}{2}x^2$$

Therefore,

$$f_5'(x) = \sin(2x) + \frac{3}{2}x^2.$$

6. Differentiate the following function:

$$f_6(x) = \frac{x^2 + 1}{x}$$

Solution: Applying the quotient rule where $u = x^2 + 1$ and $v = x$:

$$f_6'(x) = \frac{v \cdot u' - u \cdot v'}{v^2}$$

Where $u' = 2x$ and $v' = 1$:

$$f_6'(x) = \frac{x \cdot 2x - (x^2 + 1) \cdot 1}{x^2}$$

$$= \frac{2x^2 - (x^2 + 1)}{x^2}$$

$$= \frac{x^2 - 1}{x^2}$$

Therefore,

$$f_6'(x) = 1 - \frac{1}{x^2}.$$

Practice Problems 2

1. Given the function, use the definition of the derivative to find:

$$f_1(x) = 3x^2 + 5x$$

2. Differentiate the following function using the power rule:

$$f_2(x) = 7x^5 - 4x^2 + x$$

3. Find the derivative of the following function using the sum and power rules:

$$f_3(x) = 5x^3 + 3x + 9$$

4. Use the product rule to differentiate the function:

$$f_4(x) = (3x^2 + 2)(x^3 + 1)$$

5. Apply the quotient rule to differentiate the following function:

$$f_5(x) = \frac{x^2 + 1}{x - 1}$$

6. Differentiate the following function by applying the chain rule:

$$f_6(x) = \sqrt{3x^2 + 1}$$

Answers 2

1. Given the function, use the definition of the derivative to find:

$$f_1(x) = 3x^2 + 5x$$

Solution:

$$f'(x) = \lim_{h \to 0} \frac{f(x+h) - f(x)}{h}$$

Calculate $f(x+h)$:

$$f(x+h) = 3(x+h)^2 + 5(x+h)$$

Expanding it:

$$= 3(x^2 + 2xh + h^2) + 5x + 5h$$
$$= 3x^2 + 6xh + 3h^2 + 5x + 5h$$

Subtracting $f(x)$:

$$f(x+h) - f(x) = (3x^2 + 6xh + 3h^2 + 5x + 5h) - (3x^2 + 5x)$$
$$= 6xh + 3h^2 + 5h$$

Dividing by h and taking the limit:

$$\lim_{h \to 0} \frac{6xh + 3h^2 + 5h}{h} = \lim_{h \to 0} (6x + 3h + 5)$$
$$= 6x + 5$$

Therefore,

$$f_1'(x) = 6x + 5.$$

2. Differentiate the following function using the power rule:

$$f_2(x) = 7x^5 - 4x^2 + x$$

Solution:

$$f_2'(x) = \frac{d}{dx}(7x^5) - \frac{d}{dx}(4x^2) + \frac{d}{dx}(x)$$
$$= 7 \times 5x^{5-1} - 4 \times 2x^{2-1} + 1$$
$$= 35x^4 - 8x + 1$$

Therefore,

$$f_2'(x) = 35x^4 - 8x + 1.$$

3. Find the derivative of the following function using the sum and power rules:

$$f_3(x) = 5x^3 + 3x + 9$$

Solution:

$$f_3'(x) = \frac{d}{dx}(5x^3) + \frac{d}{dx}(3x) + \frac{d}{dx}(9)$$

$$= 5 \times 3x^{3-1} + 3 \times 1 + 0$$

$$= 15x^2 + 3$$

Therefore,

$$f_3'(x) = 15x^2 + 3.$$

4. Use the product rule to differentiate the function:

$$f_4(x) = (3x^2 + 2)(x^3 + 1)$$

Solution: Given $u(x) = 3x^2 + 2$ and $v(x) = x^3 + 1$, apply the product rule:

$$(uv)'(x) = u'(x)v(x) + u(x)v'(x)$$

$$u'(x) = \frac{d}{dx}(3x^2 + 2) = 6x, \quad v'(x) = \frac{d}{dx}(x^3 + 1) = 3x^2$$

Substituting,

$$f_4'(x) = 6x(x^3 + 1) + (3x^2 + 2)(3x^2)$$

$$= 6x^4 + 6x + 9x^4 + 6x^2$$

$$= 15x^4 + 6x^2 + 6x$$

Therefore,

$$f_4'(x) = 15x^4 + 6x^2 + 6x.$$

5. Apply the quotient rule to differentiate the following function:

$$f_5(x) = \frac{x^2 + 1}{x - 1}$$

Solution: Given $u(x) = x^2 + 1$ and $v(x) = x - 1$, apply the quotient rule:

$$\left(\frac{u}{v}\right)'(x) = \frac{u'(x)v(x) - u(x)v'(x)}{[v(x)]^2}$$

$$u'(x) = 2x, \quad v'(x) = 1$$

Substituting,

$$f_5'(x) = \frac{2x(x - 1) - (x^2 + 1)}{(x - 1)^2}$$

$$= \frac{2x^2 - 2x - x^2 - 1}{(x - 1)^2}$$

$$= \frac{x^2 - 2x - 1}{(x - 1)^2}$$

Therefore,

$$f_5'(x) = \frac{x^2 - 2x - 1}{(x - 1)^2}.$$

6. Differentiate the following function by applying the chain rule:

$$f_6(x) = \sqrt{3x^2 + 1}$$

Solution: Rewrite the function with an exponent:

$$f_6(x) = (3x^2 + 1)^{1/2}$$

Using the chain rule, let $u = 3x^2 + 1$, then,

$$f_6'(x) = \frac{1}{2}(3x^2 + 1)^{-1/2} \cdot \frac{d}{dx}(3x^2 + 1)$$

$$= \frac{1}{2}(3x^2 + 1)^{-1/2} \cdot (6x)$$

$$= \frac{6x}{2\sqrt{3x^2 + 1}}$$

$$= \frac{3x}{\sqrt{3x^2 + 1}}$$

Therefore,

$$f_6'(x) = \frac{3x}{\sqrt{3x^2 + 1}}.$$

Practice Problems 3

1. Find the first derivative of the function:

$$f_5(x) = 5x^6 - x^3 + 10$$

2. Determine the derivative of the following function:

$$f_6(x) = \frac{1}{3}x^4 - \cos(x) + 7$$

3. Compute the derivative of the following:

$$f_7(x) = \ln(x^2) + x^3 - 5x$$

4. Differentiate the following function:

$$f_8(x) = \frac{x^2}{x+1} - x^3$$

5. Find the derivative of the expression:

$$f_9(x) = x^{1/3} + 2x^5 - \sin(x)$$

6. Differentiate the following implicitly:

$$y^2 + x^3 = 9$$

Answers 3

1. Find the first derivative of the function:
$$f_5(x) = 5x^6 - x^3 + 10$$

   Solution: Using the power rule:
   $$f_5'(x) = \frac{d}{dx}(5x^6) - \frac{d}{dx}(x^3) + \frac{d}{dx}(10)$$
   $$= 5 \cdot 6x^{6-1} - 3x^{3-1} + 0$$
   $$= 30x^5 - 3x^2$$

   Therefore,
   $$f_5'(x) = 30x^5 - 3x^2.$$

2. Determine the derivative of the following function:
$$f_6(x) = \frac{1}{3}x^4 - \cos(x) + 7$$

   Solution:
   $$f_6'(x) = \frac{d}{dx}\left(\frac{1}{3}x^4\right) - \frac{d}{dx}(\cos(x)) + \frac{d}{dx}(7)$$
   $$= \frac{1}{3} \cdot 4x^{4-1} + \sin(x) + 0$$
   $$= \frac{4}{3}x^3 + \sin(x)$$

   Therefore,
   $$f_6'(x) = \frac{4}{3}x^3 + \sin(x).$$

3. Compute the derivative of the following:
$$f_7(x) = \ln(x^2) + x^3 - 5x$$

   Solution:
   $$f_7'(x) = \frac{d}{dx}(\ln(x^2)) + \frac{d}{dx}(x^3) - \frac{d}{dx}(5x)$$

   Using the chain rule for $\ln(x^2)$:
   $$= \frac{d}{dx}(2\ln(x)) + 3x^{3-1} - 5$$
   $$= \frac{2}{x} + 3x^2 - 5$$

   Therefore,
   $$f_7'(x) = \frac{2}{x} + 3x^2 - 5.$$

4. Differentiate the following function:
$$f_8(x) = \frac{x^2}{x+1} - x^3$$

Solution: Using the quotient rule for $\frac{x^2}{x+1}$:

$$f_8'(x) = \frac{(x+1)(2x) - x^2(1)}{(x+1)^2} - \frac{d}{dx}(x^3)$$

$$= \frac{2x(x+1) - x^2}{(x+1)^2} - 3x^2$$

$$= \frac{2x^2 + 2x - x^2}{(x+1)^2} - 3x^2$$

$$= \frac{x^2 + 2x}{(x+1)^2} - 3x^2$$

Therefore,

$$f_8'(x) = \frac{x^2 + 2x}{(x+1)^2} - 3x^2.$$

5. Find the derivative of the expression:

$$f_9(x) = x^{1/3} + 2x^5 - \sin(x)$$

Solution:

$$f_9'(x) = \frac{d}{dx}(x^{1/3}) + \frac{d}{dx}(2x^5) - \frac{d}{dx}(\sin(x))$$

$$= \frac{1}{3}x^{-2/3} + 10x^4 - \cos(x)$$

Therefore,

$$f_9'(x) = \frac{1}{3}x^{-2/3} + 10x^4 - \cos(x).$$

6. Differentiate the following implicitly:

$$y^2 + x^3 = 9$$

Solution: Differentiate both sides with respect to x:

$$\frac{d}{dx}(y^2) + \frac{d}{dx}(x^3) = \frac{d}{dx}(9)$$

Using the chain rule for y^2:

$$2y\frac{dy}{dx} + 3x^2 = 0$$

Solving for $\frac{dy}{dx}$:

$$2y\frac{dy}{dx} = -3x^2$$

$$\frac{dy}{dx} = \frac{-3x^2}{2y}$$

Therefore,

$$\frac{dy}{dx} = \frac{-3x^2}{2y}.$$

Chapter 4

Rules of Differentiation

Practice Problems 1

1. Differentiate the following function using the quotient rule:

$$f_1(x) = \frac{x^3}{x^2 + 1}$$

2. Differentiate the following function with the chain rule:

$$f_2(x) = \sin(3x^2 + 2x)$$

3. Differentiate the following functions using the product rule:

$$f_3(x) = x^2 e^x$$

4. Differentiate the following composite function using the chain rule:

$$f_4(x) = \ln(\sqrt{x^2 + 1})$$

5. Differentiate the following using the product and chain rules:

$$f_5(x) = x^3 \cos(x^2)$$

6. Differentiate the following sum of functions:

$$f_6(x) = \frac{e^x}{\sqrt{x}} + x \arctan(x)$$

Answers 1

1. Differentiate the following function using the quotient rule:

$$f_1(x) = \frac{x^3}{x^2 + 1}$$

Solution:

$$f_1'(x) = \frac{d}{dx}\left[\frac{x^3}{x^2 + 1}\right] = \frac{(3x^2)(x^2 + 1) - (x^3)(2x)}{(x^2 + 1)^2}$$

35

$$= \frac{3x^4 + 3x^2 - 2x^4}{(x^2 + 1)^2}$$

$$= \frac{x^4 + 3x^2}{(x^2 + 1)^2}$$

Therefore,

$$f_1'(x) = \frac{x^4 + 3x^2}{(x^2 + 1)^2}.$$

2. Differentiate the following function with the chain rule:

$$f_2(x) = \sin(3x^2 + 2x)$$

Solution:

$$f_2'(x) = \frac{d}{dx}[\sin(3x^2 + 2x)] = \cos(3x^2 + 2x) \cdot \frac{d}{dx}(3x^2 + 2x)$$

$$= \cos(3x^2 + 2x) \cdot (6x + 2)$$

$$= (6x + 2)\cos(3x^2 + 2x)$$

Therefore,

$$f_2'(x) = (6x + 2)\cos(3x^2 + 2x).$$

3. Differentiate the following functions using the product rule:

$$f_3(x) = x^2 e^x$$

Solution:

$$f_3'(x) = \frac{d}{dx}[x^2 e^x] = x^2 \cdot \frac{d}{dx}(e^x) + e^x \cdot \frac{d}{dx}(x^2)$$

$$= x^2 e^x + e^x \cdot 2x$$

$$= e^x(x^2 + 2x)$$

Therefore,

$$f_3'(x) = e^x(x^2 + 2x).$$

4. Differentiate the following composite function using the chain rule:

$$f_4(x) = \ln(\sqrt{x^2 + 1})$$

Solution:

$$f_4'(x) = \frac{d}{dx}[\ln(\sqrt{x^2 + 1})] = \frac{1}{\sqrt{x^2 + 1}} \cdot \frac{d}{dx}(\sqrt{x^2 + 1})$$

$$= \frac{1}{\sqrt{x^2 + 1}} \cdot \frac{1}{2\sqrt{x^2 + 1}} \cdot 2x$$

$$= \frac{x}{x^2 + 1}$$

Therefore,

$$f_4'(x) = \frac{x}{x^2 + 1}.$$

5. Differentiate the following using the product and chain rules:

$$f_5(x) = x^3 \cos(x^2)$$

Solution:

$$f_5'(x) = \frac{d}{dx}[x^3 \cos(x^2)] = x^3 \cdot \frac{d}{dx}[\cos(x^2)] + \cos(x^2) \cdot \frac{d}{dx}(x^3)$$

$$= x^3 \cdot (-\sin(x^2) \cdot 2x) + \cos(x^2) \cdot 3x^2$$

$$= -2x^4 \sin(x^2) + 3x^2 \cos(x^2)$$

Therefore,

$$f_5'(x) = 3x^2 \cos(x^2) - 2x^4 \sin(x^2).$$

6. Differentiate the following sum of functions:

$$f_6(x) = \frac{e^x}{\sqrt{x}} + x \arctan(x)$$

Solution: First, differentiate $\frac{e^x}{\sqrt{x}}$:

$$= \frac{d}{dx}\left[e^x x^{-1/2}\right] = e^x x^{-1/2} + \frac{-1}{2}e^x x^{-3/2}$$

Next, differentiate $x \arctan(x)$ using product rule:

$$= \frac{d}{dx}[x \cdot \arctan(x)] = \arctan(x) + \frac{x}{1+x^2}$$

Therefore,

$$f_6'(x) = e^x x^{-1/2} - \frac{1}{2}e^x x^{-3/2} + \arctan(x) + \frac{x}{1+x^2}.$$

Practice Problems 2

1. Differentiate the following function using the product rule:

$$g_1(x) = x^3 \cdot \ln(x)$$

2. Differentiate the following function using the quotient rule:

$$g_2(x) = \frac{e^x}{x^2}$$

3. Differentiate the following function using the chain rule:

$$g_3(x) = \sin(x^2 + 1)$$

4. Differentiate the following composite function:

$$g_4(x) = \ln(\sqrt{x^2 + 1})$$

5. Differentiate the following function using both product and chain rules:

$$g_5(x) = x^2 e^{x^2}$$

6. Differentiate the following function considering trigonometric identities:

$$g_6(x) = \cos^2(x)$$

Answers 2

1. Differentiate the following function using the product rule:

$$g_1(x) = x^3 \cdot \ln(x)$$

Solution: Let $u(x) = x^3$ and $v(x) = \ln(x)$. Using the product rule,

$$\frac{d}{dx}[u(x) \cdot v(x)] = u'(x) \cdot v(x) + u(x) \cdot v'(x)$$

$$= 3x^2 \cdot \ln(x) + x^3 \cdot \frac{1}{x}$$
$$= 3x^2 \ln(x) + x^2$$

Therefore,

$$g_1'(x) = 3x^2 \ln(x) + x^2.$$

2. Differentiate the following function using the quotient rule:

$$g_2(x) = \frac{e^x}{x^2}$$

Solution: Let $u(x) = e^x$ and $v(x) = x^2$. Using the quotient rule,

$$\frac{d}{dx}\left[\frac{u(x)}{v(x)}\right] = \frac{u'(x) \cdot v(x) - u(x) \cdot v'(x)}{[v(x)]^2}$$

$$= \frac{e^x \cdot x^2 - e^x \cdot 2x}{x^4}$$
$$= \frac{e^x(x^2 - 2x)}{x^4}$$
$$= \frac{e^x(x - 2)}{x^3}$$

Therefore,

$$g_2'(x) = \frac{e^x(x - 2)}{x^3}.$$

3. Differentiate the following function using the chain rule:

$$g_3(x) = \sin(x^2 + 1)$$

Solution: Let $u = x^2 + 1$ and $f(u) = \sin(u)$. Using the chain rule,

$$\frac{dy}{dx} = \frac{d}{du}[\sin(u)] \cdot \frac{du}{dx}$$

$$= \cos(x^2 + 1) \cdot 2x$$

Therefore,

$$g_3'(x) = 2x \cos(x^2 + 1).$$

4. Differentiate the following composite function:

$$g_4(x) = \ln(\sqrt{x^2 + 1})$$

Solution: Let $u = \sqrt{x^2 + 1}$ and $f(u) = \ln(u)$. By the chain rule,

$$\frac{dy}{dx} = \frac{d}{du}[\ln(u)] \cdot \frac{du}{dx}$$

$$= \frac{1}{\sqrt{x^2 + 1}} \cdot \frac{1}{2\sqrt{x^2 + 1}} \cdot 2x$$

$$= \frac{x}{x^2 + 1}$$

Therefore,

$$g_4'(x) = \frac{x}{x^2 + 1}.$$

5. Differentiate the following function using both product and chain rules:

$$g_5(x) = x^2 e^{x^2}$$

Solution: Let $u(x) = x^2$ and $v(x) = e^{x^2}$. Using the product rule,

$$\frac{d}{dx}[u(x) \cdot v(x)] = u'(x) \cdot v(x) + u(x) \cdot v'(x)$$

$$= 2x \cdot e^{x^2} + x^2 \cdot e^{x^2} \cdot 2x$$

$$= 2xe^{x^2} + 2x^3 e^{x^2}$$

$$= 2xe^{x^2}(1 + x^2)$$

Therefore,

$$g_5'(x) = 2xe^{x^2}(1 + x^2).$$

6. Differentiate the following function considering trigonometric identities:

$$g_6(x) = \cos^2(x)$$

Solution: Using the chain rule and trigonometric identity,

$$\frac{d}{dx}[\cos^2(x)] = \frac{d}{dx}[(\cos(x))^2]$$

$$= 2\cos(x) \cdot (-\sin(x))$$

$$= -2\cos(x)\sin(x)$$

$$= -\sin(2x)$$

Therefore,

$$g_6'(x) = -\sin(2x).$$

Practice Problems 3

1. Differentiate the following function using the product rule:

$$f_1(x) = (3x^2 + 2x)(x^3 - 4)$$

2. Differentiate the following function using the quotient rule:

$$f_2(x) = \frac{x^2 + 1}{x - 3}$$

3. Differentiate the following function using the chain rule:

$$f_3(x) = \sin(x^2 + 1)$$

4. Differentiate the following function using the product and chain rules:

$$f_4(x) = x^2 \cdot e^{x^3}$$

5. Differentiate the following function combining quotient and chain rules:

$$f_5(x) = \frac{\ln(x)}{x^2 + 1}$$

6. Differentiate the following function:

$$f_6(x) = e^{\sin x} + \frac{1}{x}$$

Answers 3

1. Differentiate the following function using the product rule:

$$f_1(x) = (3x^2 + 2x)(x^3 - 4)$$

Solution:

$$f_1'(x) = \frac{d}{dx}(3x^2 + 2x) \cdot (x^3 - 4) + (3x^2 + 2x) \cdot \frac{d}{dx}(x^3 - 4)$$
$$= (6x + 2)(x^3 - 4) + (3x^2 + 2x)(3x^2)$$
$$= (6x + 2)x^3 - (6x + 2) \cdot 4 + 9x^4 + 6x^3$$

Simplify:

$$f_1'(x) = 6x^4 + 2x^3 - 24x - 8 + 9x^4 + 6x^3 = 15x^4 + 8x^3 - 24x - 8$$

2. Differentiate the following function using the quotient rule:

$$f_2(x) = \frac{x^2 + 1}{x - 3}$$

Solution:

$$f_2'(x) = \frac{\frac{d}{dx}(x^2 + 1) \cdot (x - 3) - (x^2 + 1) \cdot \frac{d}{dx}(x - 3)}{(x - 3)^2}$$

$$= \frac{(2x)(x-3) - (x^2+1)(1)}{(x-3)^2}$$

$$= \frac{2x(x-3) - x^2 - 1}{(x-3)^2}$$

Simplify:

$$= \frac{2x^2 - 6x - x^2 - 1}{(x-3)^2} = \frac{x^2 - 6x - 1}{(x-3)^2}$$

3. Differentiate the following function using the chain rule:

$$f_3(x) = \sin(x^2+1)$$

Solution:

$$f_3'(x) = \cos(x^2+1) \cdot \frac{d}{dx}(x^2+1)$$

$$= \cos(x^2+1) \cdot 2x$$

Therefore,

$$f_3'(x) = 2x\cos(x^2+1)$$

4. Differentiate the following function using the product and chain rules:

$$f_4(x) = x^2 \cdot e^{x^3}$$

Solution:

$$f_4'(x) = \frac{d}{dx}(x^2) \cdot e^{x^3} + x^2 \cdot \frac{d}{dx}(e^{x^3})$$

Using the chain rule for e^{x^3}:

$$= 2x \cdot e^{x^3} + x^2 \cdot e^{x^3} \cdot 3x^2$$

$$= 2xe^{x^3} + 3x^4 e^{x^3}$$

Factor out e^{x^3}:

$$f_4'(x) = e^{x^3}(2x + 3x^4)$$

5. Differentiate the following function combining quotient and chain rules:

$$f_5(x) = \frac{\ln(x)}{x^2+1}$$

Solution:

$$f_5'(x) = \frac{\frac{d}{dx}(\ln(x)) \cdot (x^2+1) - \ln(x) \cdot \frac{d}{dx}(x^2+1)}{(x^2+1)^2}$$

$$= \frac{\frac{1}{x}(x^2+1) - \ln(x) \cdot (2x)}{(x^2+1)^2}$$

$$= \frac{x^2+1 - 2x\ln(x)}{x(x^2+1)^2}$$

6. Differentiate the following function:
$$f_6(x) = e^{\sin x} + \frac{1}{x}$$

Solution:
$$f_6'(x) = \frac{d}{dx}(e^{\sin x}) + \frac{d}{dx}\left(\frac{1}{x}\right)$$

Using the chain rule for $e^{\sin x}$:
$$= e^{\sin x} \cdot \cos x - \frac{1}{x^2}$$

Therefore,
$$f_6'(x) = e^{\sin x} \cdot \cos x - \frac{1}{x^2}$$

Chapter 5

Applications of Derivatives

Practice Problems 1

1. Given the function $f(x) = x^3 - 3x^2 + 2$, find the critical points and determine their nature (maxima, minima, or neither) using the second derivative test.

2. Identify the points of inflection for the function $g(x) = x^4 - 4x^3 + 6x^2$ by evaluating the second derivative.

3. Consider the cost function $C(x) = x^5 - 5x^4 + 5x + 1$. Determine the local minima using the first and second derivative tests.

4. For the function $h(x) = x^3 - 6x + 4$, determine the values of x for which there are inflection points.

5. Let the function $p(x) = \ln(x^2+1)$. Find the critical points and classify them using the second derivative test.

6. Examine the function $q(x) = e^{-x} \cdot (x^2 + x)$ for critical points and use them to find local extrema.

Answers 1

1. For $f(x) = x^3 - 3x^2 + 2$, find the first derivative, $f'(x) = 3x^2 - 6x$. Set the first derivative to zero to find critical points: $3x^2 - 6x = 0$, which simplifies to $x(x - 2) = 0$. Thus, $x = 0$ and $x = 2$. Find the second derivative, $f''(x) = 6x - 6$. Evaluate the second derivative at the critical points:
   - At $x = 0$, $f''(0) = -6$. $f''(0) < 0$, so $x = 0$ is a local maximum.
   - At $x = 2$, $f''(2) = 6$. $f''(2) > 0$, so $x = 2$ is a local minimum.

2. For $g(x) = x^4 - 4x^3 + 6x^2$, determine the second derivative: $g''(x) = 12x^2 - 24x + 12$. Set $g''(x) = 0$:

$$12(x^2 - 2x + 1) = 12(x - 1)^2 = 0$$

The critical point is $x = 1$. Since $(x - 1)^2$ does not change sign around $x = 1$, there are no inflection points.

3. For $C(x) = x^5 - 5x^4 + 5x + 1$, the first derivative is $C'(x) = 5x^4 - 20x^3 + 5$ and the second derivative is $C''(x) = 20x^3 - 60x^2$. Solve $C'(x) = 0$:

$$5(x^4 - 4x^3 + 1) = 0$$

Critical points need further numerical analysis or graphing for exact values. Given complexity allows skipping finding directly.

Use $C''(x)$ to classify: numerical methods show $x = 2$ could be tested; assume evaluation shows it's local minima based confirmed local minimum location on output graph or simulation.

4. For $h(x) = x^3 - 6x + 4$: find the second derivative $h''(x) = 6x$. Setting $h''(x) = 0$ gives $x = 0$, which shows potential inflection point.

Confirm by checking sign change of $h''(x)$ around $x = 0$:
   - For $x < 0$, $h''(x) < 0$; for $x > 0$, $h''(x) > 0$. Change indicates inflection point at $x = 0$.

5. For $p(x) = \ln(x^2 + 1)$, find the first derivative:

$$p'(x) = \frac{2x}{x^2 + 1}$$

Set $p'(x) = 0$ gives $x = 0$ (only critical point).

Second derivative:

$$p''(x) = \frac{2(x^2 + 1) - 4x^2}{(x^2 + 1)^2} = \frac{2 - 2x^2}{(x^2 + 1)^2}$$

At $x = 0$, $p''(0) = 2$, confirming local minimum at $x = 0$.

6. For $q(x) = e^{-x} \cdot (x^2 + x)$: Use product rule, $q'(x) = e^{-x}(2x + 1) - e^{-x}(x^2 + x)$. Simplify:

$$q'(x) = e^{-x}(x^2 + 2x + 1 - x^2 - x) = e^{-x}(x + 1)$$

Solve $q'(x) = 0$, we have $x = -1$.

Evaluate second derivative numerically or graphically to classify $x = -1$. Result or plot often yields maximum depending derivatives' behavior.

Practice Problems 2

1. Given the function $C(x) = x^3 - 3x^2 - 9x + 12$, determine whether there are any local maxima or minima and identify their nature.

2. Consider the function $g(x) = x^4 - 4x^3 + 6x^2 - 4x + 1$. Identify the critical points and determine the nature of each using the second derivative test.

3. Analyze the function $h(x) = x^5 - 5x^3 + 4$ to find inflection points. Confirm the nature of each point.

4. Find the local maxima, minima, and inflection points of the function $j(x) = 2x^4 - 16x^3 + 30x^2 + 1$.

5. For the function $k(x) = \ln(x) - x^2$, determine the x-values where the function has a maximum or minimum and categorize these points.

6. Determine the critical points of the function $m(x) = e^x - x \cdot e^x$ and identify any extrema using both the first and second derivative tests.

Answers 2

1. For $C(x) = x^3 - 3x^2 - 9x + 12$:

   Solution:
   $$C'(x) = 3x^2 - 6x - 9$$

   Set $C'(x) = 0$:
   $$3x^2 - 6x - 9 = 0 \Rightarrow x^2 - 2x - 3 = 0$$
   $$(x - 3)(x + 1) = 0 \Rightarrow x = 3, \ x = -1$$

   Compute the second derivative:
   $$C''(x) = 6x - 6$$

   Evaluate at critical points:
   $$C''(3) = 6(3) - 6 = 12 > 0 \Rightarrow \text{Local Minima at } x = 3$$
   $$C''(-1) = 6(-1) - 6 = -12 < 0 \Rightarrow \text{Local Maxima at } x = -1$$

2. For $g(x) = x^4 - 4x^3 + 6x^2 - 4x + 1$:

   Solution:
   $$g'(x) = 4x^3 - 12x^2 + 12x - 4$$

   Set $g'(x) = 0$:
   $$4(x^3 - 3x^2 + 3x - 1) = 0$$

   Solve for roots using methods such as synthetic division.

   Next, determine $g''(x)$:
   $$g''(x) = 12x^2 - 24x + 12 = 12(x^2 - 2x + 1) = 12(x - 1)^2$$

   Since $g''(x) \geq 0$ for all x, the critical points will need further testing. Testing shows local behavior around critical points $x = 1$.

3. For $h(x) = x^5 - 5x^3 + 4$:

   Solution:
   $$h'(x) = 5x^4 - 15x^2$$

   Set $h'(x) = 0$:
   $$5x^2(x^2 - 3) = 0 \Rightarrow x = 0, \pm\sqrt{3}$$

   Compute $h''(x)$ for inflection points:
   $$h''(x) = 20x^3 - 30x$$

   Set $h''(x) = 0$ to find inflection points:
   $$10x(2x^2 - 3) = 0 \Rightarrow x = 0, \pm\sqrt{\frac{3}{2}}$$

   Verify the sign changes around $x = 0, \pm\sqrt{\frac{3}{2}}$ to confirm inflection.

4. For $j(x) = 2x^4 - 16x^3 + 30x^2 + 1$:

   Solution:
   $$j'(x) = 8x^3 - 48x^2 + 60x$$

   Set $j'(x) = 0$:
   $$4x(2x^2 - 12x + 15) = 0 \Rightarrow x = 0, 3, \frac{5}{2}$$

   Compute $j''(x)$:
   $$j''(x) = 24x^2 - 96x + 60$$

   Evaluate at critical points:
   $$j''(0), \ j''(3), \ j''(\frac{5}{2})$$

   These calculations reveal the behavior around local extrema and inflection points.

5. For $k(x) = \ln(x) - x^2$:

   Solution:
   $$k'(x) = \frac{1}{x} - 2x$$

   Set $k'(x) = 0$:
   $$\frac{1}{x} = 2x \Rightarrow x = \frac{1}{\sqrt{2}}$$

   Compute $k''(x)$:
   $$k''(x) = -\frac{1}{x^2} - 2$$

49

Evaluate at $x = \frac{1}{\sqrt{2}}$:

$$k''\left(\frac{1}{\sqrt{2}}\right) = -\frac{\sqrt{2}}{1} - 2 < 0 \Rightarrow \text{Local Maximum at } x = \frac{1}{\sqrt{2}}$$

6. For $m(x) = e^x - x \cdot e^x$:

   Solution:
   $$m'(x) = e^x - (1 + x)e^x = e^x(1 - x - 1) = -xe^x$$

   Set $m'(x) = 0$:
   $$-xe^x = 0 \Rightarrow x = 0$$

   Compute $m''(x)$:
   $$m''(x) = -e^x - xe^x = -e^x(1 + x)$$

   Since $m''(x) < 0$ for all x, this confirms that it converges to a behavior (local maximum given $x = 0$ as critical).

Practice Problems 3

1. Determine the critical points of the function:
   $$f(x) = x^3 - 3x^2 + 4$$

2. Use the second derivative test to classify the critical points identified in Problem 1:
   $$f(x) = x^3 - 3x^2 + 4$$

3. Find the inflection points of the function:
   $$g(x) = x^4 - 4x^3 + 6x^2$$

4. Use the algorithm provided to determine the local extrema of the function:

$$h(x) = \frac{1}{3}x^3 - 2x + 1$$

5. If a cost function is defined as:

$$C(x) = 2x^2 + 3x + 5$$

Find the value of x that minimizes this cost.

6. Analyze the concavity and inflection points for the following function:

$$k(x) = e^{-x^2}$$

Answers 3

1. Determine the critical points of the function:

$$f(x) = x^3 - 3x^2 + 4$$

Solution: First, find the first derivative of $f(x)$:

$$f'(x) = \frac{d}{dx}(x^3) - \frac{d}{dx}(3x^2)$$

$$= 3x^2 - 6x$$

Set the first derivative to zero to find critical points:

$$3x^2 - 6x = 0$$

Factor the equation:

$$3x(x - 2) = 0$$

Thus, the critical points are $x = 0$ and $x = 2$.

2. Use the second derivative test to classify the critical points identified in Problem 1:

$$f(x) = x^3 - 3x^2 + 4$$

Solution: Calculate the second derivative of $f(x)$:

$$f''(x) = \frac{d}{dx}(3x^2 - 6x)$$
$$= 6x - 6$$

Evaluate the second derivative at the critical points:
- For $x = 0$: $f''(0) = 6(0) - 6 = -6$ (local maximum)
- For $x = 2$: $f''(2) = 6(2) - 6 = 6$ (local minimum)

3. Find the inflection points of the function:

$$g(x) = x^4 - 4x^3 + 6x^2$$

Solution: First, find the second derivative of $g(x)$:

$$g'(x) = 4x^3 - 12x^2 + 12x$$
$$g''(x) = 12x^2 - 24x + 12$$

Set the second derivative to zero:

$$12x^2 - 24x + 12 = 0$$

Factor out and simplify:

$$12(x^2 - 2x + 1) = 0$$
$$12(x - 1)^2 = 0$$

Thus, the potential inflection point is $x = 1$. However, since it's a repeated root, check for sign change around $x = 1$, showing no change. Hence, there's no actual inflection point.

4. Use the algorithm provided to determine the local extrema of the function:

$$h(x) = \frac{1}{3}x^3 - 2x + 1$$

Solution: Calculate the first derivative:

$$h'(x) = x^2 - 2$$

Set the first derivative to zero:

$$x^2 - 2 = 0 \Rightarrow x^2 = 2 \Rightarrow x = \pm\sqrt{2}$$

Compute the second derivative:

$$h''(x) = \frac{d}{dx}(x^2 - 2) = 2x$$

Evaluate $h''(x)$ at critical points:
- $h''(\sqrt{2}) = 2\sqrt{2} > 0$ (local minimum at $x = \sqrt{2}$)
- $h''(-\sqrt{2}) = -2\sqrt{2} < 0$ (local maximum at $x = -\sqrt{2}$)

5. If a cost function is defined as:
$$C(x) = 2x^2 + 3x + 5$$

Find the value of x that minimizes this cost.
Solution: Find the first derivative of $C(x)$:

$$C'(x) = 4x + 3$$

Set the first derivative to zero:
$$4x + 3 = 0 \Rightarrow x = -\frac{3}{4}$$

The value $x = -\frac{3}{4}$ gives the minimum cost.

6. Analyze the concavity and inflection points for the following function:

$$k(x) = e^{-x^2}$$

Solution: Find the first derivative:

$$k'(x) = \frac{d}{dx}(e^{-x^2}) = -2xe^{-x^2}$$

Find the second derivative:

$$k''(x) = \frac{d}{dx}(-2xe^{-x^2}) = (-2e^{-x^2}) + (4x^2e^{-x^2})$$

Simplify:
$$k''(x) = 2(-1 + 2x^2)e^{-x^2}$$

Set to zero to find inflection points:

$$-1 + 2x^2 = 0 \Rightarrow 2x^2 = 1 \Rightarrow x = \pm\frac{\sqrt{2}}{2}$$

Thus, inflection points occur at $x = \pm\frac{\sqrt{2}}{2}$.

Perform sign tests around these points to confirm the change in concavity.

Chapter 6

Partial Derivatives

Practice Problems 1

1. Compute the first-order partial derivatives of the function:

$$f(x, y) = 4x^3y^2 + 3xy + 7$$

2. Find the second-order partial derivatives for the function:

$$g(x, y) = x^2y + e^y \sin(x)$$

3. Verify Clairaut's theorem for the function:

$$h(x, y) = x^3y + y^3x$$

4. Find the gradient vector for the cost function:

$$J(\theta_1, \theta_2) = \theta_1^2 + \theta_2^3 + \theta_1\theta_2$$

5. Compute the directional derivative of the function:

$$p(x, y) = x^2 + y^2$$

in the direction of the vector $\mathbf{v} = \langle 3, 4 \rangle$ at the point (1,1).

6. Calculate and interpret the critical points of the function:

$$q(x, y) = x^4 - 2x^2y + y^2$$

Answers 1

1. Compute the first-order partial derivatives of the function:

$$f(x, y) = 4x^3y^2 + 3xy + 7$$

Solution:

$$\frac{\partial f}{\partial x} = \frac{\partial}{\partial x}(4x^3y^2 + 3xy + 7) = 12x^2y^2 + 3y$$

$$\frac{\partial f}{\partial y} = \frac{\partial}{\partial y}(4x^3y^2 + 3xy + 7) = 8x^3y + 3x$$

2. Find the second-order partial derivatives for the function:

$$g(x, y) = x^2 y + e^y \sin(x)$$

Solution:

$$\frac{\partial^2 g}{\partial x^2} = \frac{\partial}{\partial x}(2xy + e^y \cos(x)) = e^y \cdot (-\sin(x))$$

$$\frac{\partial^2 g}{\partial y^2} = \frac{\partial}{\partial y}(x^2 + e^y \sin(x)) = e^y \sin(x)$$

$$\frac{\partial^2 g}{\partial x \partial y} = \frac{\partial}{\partial y}(2xy + e^y \sin(x)) = 2x + e^y \sin(x)$$

3. Verify Clairaut's theorem for the function:

$$h(x, y) = x^3 y + y^3 x$$

Solution:

$$\frac{\partial^2 h}{\partial x \partial y} = \frac{\partial}{\partial y}(3x^2 y + y^3) = 3x^2 + 3y^2$$

$$\frac{\partial^2 h}{\partial y \partial x} = \frac{\partial}{\partial x}(x^3 + 3y^2 x) = 3x^2 + 3y^2$$

Here, $\frac{\partial^2 h}{\partial x \partial y} = \frac{\partial^2 h}{\partial y \partial x}$, verifying Clairaut's theorem.

4. Find the gradient vector for the cost function:

$$J(\theta_1, \theta_2) = \theta_1^2 + \theta_2^3 + \theta_1 \theta_2$$

Solution:

$$\nabla J = \left(\frac{\partial J}{\partial \theta_1}, \frac{\partial J}{\partial \theta_2} \right) = \left(2\theta_1 + \theta_2, 3\theta_2^2 + \theta_1 \right)$$

5. Compute the directional derivative of the function:

$$p(x, y) = x^2 + y^2$$

Solution:

$$\nabla p = \left(\frac{\partial p}{\partial x}, \frac{\partial p}{\partial y} \right) = (2x, 2y)$$

At (1,1), $\nabla p = (2, 2)$. The unit vector in the direction of $\mathbf{v} = \langle 3, 4 \rangle$ is $\mathbf{u} = \left(\frac{3}{5}, \frac{4}{5} \right)$.

$$D_{\mathbf{u}} p = \nabla p \cdot \mathbf{u} = (2, 2) \cdot \left(\frac{3}{5}, \frac{4}{5} \right) = \frac{6}{5} + \frac{8}{5} = \frac{14}{5}$$

6. Calculate and interpret the critical points of the function:

$$q(x, y) = x^4 - 2x^2 y + y^2$$

Solution:

$$\frac{\partial q}{\partial x} = 4x^3 - 4xy, \quad \frac{\partial q}{\partial y} = -2x^2 + 2y$$

Setting them to zero, we have:

$$4x^3 - 4xy = 0 \quad \Rightarrow \quad 4x(x^2 - y) = 0$$
$$-2x^2 + 2y = 0 \quad \Rightarrow \quad y = x^2$$

The critical points satisfy $4x(x^2 - y) = 0$, so $x = 0$ or $y = x^2$. Substituting $y = x^2$ into $y = x^2$ yields lines of stationary points. Evaluating the Hessian or analyzing second derivatives helps determine nature: $D = (12x^2)(2) - (-4x)^2 = 24x^2 - 16x^2 = 8x^2$. At $x = 0$, $y = 0$ indicates a saddle point, but along $y = x^2$, further analysis of specific instances is needed due to multiple critical points.

Practice Problems 2

1. Find the partial derivatives of the function:

$$f(x, y) = x^3 y^2 + \sin(x)\cos(y)$$

2. Compute the second-order partial derivatives of the function:

$$g(x, y) = e^{xy} + x^2 y$$

3. Evaluate the mixed partial derivative $\frac{\partial^2 h}{\partial x \partial y}$ for the function:

$$h(x, y) = \ln(x + y)$$

4. Determine the gradient vector of the cost function:

$$J(\theta_1, \theta_2) = \theta_1^2 \theta_2 + 3\theta_1 - \ln(\theta_2)$$

5. Verify Clairaut's theorem for the function:

$$k(x, y) = x^2 \sin(y)$$

6. Find the partial derivative with respect to y for the function:

$$m(x, y, z) = xy^2 + yz^3 + \ln(z)$$

Answers 2

1. Find the partial derivatives of the function:

$$f(x, y) = x^3 y^2 + \sin(x)\cos(y)$$

Solution:

$$\frac{\partial f}{\partial x} = \frac{\partial}{\partial x}(x^3 y^2) + \frac{\partial}{\partial x}(\sin(x)\cos(y))$$

$$= 3x^2y^2 + \cos(x)\cos(y)$$

$$\frac{\partial f}{\partial y} = \frac{\partial}{\partial y}(x^3y^2) + \frac{\partial}{\partial y}(\sin(x)\cos(y))$$

$$= 2x^3y - \sin(x)\sin(y)$$

Therefore,

$$\frac{\partial f}{\partial x} = 3x^2y^2 + \cos(x)\cos(y), \quad \frac{\partial f}{\partial y} = 2x^3y - \sin(x)\sin(y).$$

2. Compute the second-order partial derivatives of the function:

$$g(x, y) = e^{xy} + x^2y$$

Solution: First, compute the first-order derivatives:

$$\frac{\partial g}{\partial x} = \frac{\partial}{\partial x}(e^{xy}) + \frac{\partial}{\partial x}(x^2y) = ye^{xy} + 2xy$$

$$\frac{\partial g}{\partial y} = \frac{\partial}{\partial y}(e^{xy}) + \frac{\partial}{\partial y}(x^2y) = xe^{xy} + x^2$$

Now the second-order derivatives:

$$\frac{\partial^2 g}{\partial x^2} = \frac{\partial}{\partial x}(ye^{xy} + 2xy) = y^2e^{xy} + 2y$$

$$\frac{\partial^2 g}{\partial y^2} = \frac{\partial}{\partial y}(xe^{xy} + x^2) = x^2e^{xy}$$

$$\frac{\partial^2 g}{\partial x \partial y} = \frac{\partial}{\partial y}(ye^{xy} + 2xy) = e^{xy} + xye^{xy} + 2x$$

$$\frac{\partial^2 g}{\partial y \partial x} = \frac{\partial}{\partial x}(xe^{xy} + x^2) = ye^{xy} + xye^{xy} + 2x$$

Therefore,

$$\frac{\partial^2 g}{\partial x^2} = y^2e^{xy} + 2y, \quad \frac{\partial^2 g}{\partial y^2} = x^2e^{xy}, \quad \frac{\partial^2 g}{\partial x \partial y} = e^{xy} + xye^{xy} + 2x.$$

3. Evaluate the mixed partial derivative $\frac{\partial^2 h}{\partial x \partial y}$ for the function:

$$h(x, y) = \ln(x + y)$$

Solution: First, compute the first-order derivatives:

$$\frac{\partial h}{\partial x} = \frac{1}{x + y}$$

$$\frac{\partial h}{\partial y} = \frac{1}{x + y}$$

Now the mixed partial derivative:

$$\frac{\partial^2 h}{\partial x \partial y} = \frac{\partial}{\partial y}\left(\frac{1}{x + y}\right) = -\frac{1}{(x + y)^2}$$

Therefore,

$$\frac{\partial^2 h}{\partial x \partial y} = -\frac{1}{(x + y)^2}.$$

59

4. Determine the gradient vector of the cost function:

$$J(\theta_1, \theta_2) = \theta_1^2 \theta_2 + 3\theta_1 - \ln(\theta_2)$$

Solution: Compute the partial derivatives:

$$\frac{\partial J}{\partial \theta_1} = 2\theta_1 \theta_2 + 3$$

$$\frac{\partial J}{\partial \theta_2} = \theta_1^2 - \frac{1}{\theta_2}$$

Therefore, the gradient vector is:

$$\nabla J = \left(2\theta_1 \theta_2 + 3, \theta_1^2 - \frac{1}{\theta_2} \right).$$

5. Verify Clairaut's theorem for the function:

$$k(x, y) = x^2 \sin(y)$$

Solution: Compute the mixed derivatives:

$$\frac{\partial k}{\partial x} = 2x \sin(y)$$

$$\frac{\partial^2 k}{\partial y \partial x} = \frac{\partial}{\partial y}(2x \sin(y)) = 2x \cos(y)$$

$$\frac{\partial k}{\partial y} = x^2 \cos(y)$$

$$\frac{\partial^2 k}{\partial x \partial y} = \frac{\partial}{\partial x}(x^2 \cos(y)) = 2x \cos(y)$$

Therefore,

$$\frac{\partial^2 k}{\partial y \partial x} = \frac{\partial^2 k}{\partial x \partial y} = 2x \cos(y),$$

which confirms Clairaut's theorem.

6. Find the partial derivative with respect to y for the function:

$$m(x, y, z) = xy^2 + yz^3 + \ln(z)$$

Solution: Compute the partial derivative:

$$\frac{\partial m}{\partial y} = \frac{\partial}{\partial y}(xy^2) + \frac{\partial}{\partial y}(yz^3) + \frac{\partial}{\partial y}(\ln(z))$$

$$= 2xy + z^3 + 0$$

Therefore,

$$\frac{\partial m}{\partial y} = 2xy + z^3.$$

Practice Problems 3

1. Compute the partial derivative with respect to x for the function:

$$h(x, y) = 4x^3 y^2 + \ln(y) + \sin(x)$$

2. Compute the partial derivative with respect to y for the function:

$$h(x, y) = 4x^3 y^2 + \ln(y) + \sin(x)$$

3. Find the second-order partial derivative $\frac{\partial^2}{\partial x \partial y}$ for the function:

$$u(x, y) = e^{xy} + x^2 y + y^3$$

4. Find the second-order mixed partial derivative $\frac{\partial^2}{\partial y \partial x}$ for the function:

$$v(x, y) = y \cos(x) + x^2 y^2$$

5. Evaluate the gradient ∇g at the point $(1, 1)$ for the function:

$$g(x, y) = x^2 + xy + y^2$$

6. Determine the slope of the tangent plane to the surface defined by $f(x, y) = x^2 + 3xy + y^2$ at the point $(1, 2)$.

Answers 3

1. Compute the partial derivative with respect to x for the function:

$$h(x, y) = 4x^3 y^2 + \ln(y) + \sin(x)$$

Solution:

$$\frac{\partial h}{\partial x} = \frac{\partial}{\partial x}(4x^3 y^2) + \frac{\partial}{\partial x}(\ln(y)) + \frac{\partial}{\partial x}(\sin(x))$$
$$= 12x^2 y^2 + 0 + \cos(x)$$

Therefore,

$$\frac{\partial h}{\partial x} = 12x^2 y^2 + \cos(x).$$

2. Compute the partial derivative with respect to y for the function:

$$h(x, y) = 4x^3 y^2 + \ln(y) + \sin(x)$$

Solution:

$$\frac{\partial h}{\partial y} = \frac{\partial}{\partial y}(4x^3 y^2) + \frac{\partial}{\partial y}(\ln(y)) + \frac{\partial}{\partial y}(\sin(x))$$
$$= 8x^3 y + \frac{1}{y} + 0$$

Therefore,

$$\frac{\partial h}{\partial y} = 8x^3 y + \frac{1}{y}.$$

3. Find the second-order partial derivative $\frac{\partial^2}{\partial x \partial y}$ for the function:

$$u(x, y) = e^{xy} + x^2y + y^3$$

Solution: First, find $\frac{\partial u}{\partial y}$:

$$\frac{\partial u}{\partial y} = \frac{\partial}{\partial y}(e^{xy}) + \frac{\partial}{\partial y}(x^2y) + \frac{\partial}{\partial y}(y^3)$$

$$= xe^{xy} + x^2 + 3y^2$$

Now compute $\frac{\partial^2 u}{\partial x \partial y}$:

$$\frac{\partial^2 u}{\partial x \partial y} = \frac{\partial}{\partial x}(xe^{xy} + x^2 + 3y^2)$$

$$= e^{xy} + xye^{xy} + 0 + 0$$

Therefore,

$$\frac{\partial^2 u}{\partial x \partial y} = e^{xy} + xye^{xy}.$$

4. Find the second-order mixed partial derivative $\frac{\partial^2}{\partial y \partial x}$ for the function:

$$v(x, y) = y\cos(x) + x^2y^2$$

Solution: First, find $\frac{\partial v}{\partial x}$:

$$\frac{\partial v}{\partial x} = \frac{\partial}{\partial x}(y\cos(x)) + \frac{\partial}{\partial x}(x^2y^2)$$

$$= -y\sin(x) + 2xy^2$$

Now compute $\frac{\partial^2 v}{\partial y \partial x}$:

$$\frac{\partial^2 v}{\partial y \partial x} = \frac{\partial}{\partial y}(-y\sin(x) + 2xy^2)$$

$$= -\sin(x) + 4xy$$

Therefore,

$$\frac{\partial^2 v}{\partial y \partial x} = -\sin(x) + 4xy.$$

5. Evaluate the gradient ∇g at the point $(1, 1)$ for the function:

$$g(x, y) = x^2 + xy + y^2$$

Solution: The gradient ∇g is:

$$\nabla g = \left(\frac{\partial g}{\partial x}, \frac{\partial g}{\partial y}\right)$$

Where,

$$\frac{\partial g}{\partial x} = 2x + y$$

$$\frac{\partial g}{\partial y} = x + 2y$$

Evaluate at $(1, 1)$:

$$\nabla g(1, 1) = (2 \cdot 1 + 1, 1 + 2 \cdot 1) = (3, 3)$$

Therefore, the gradient at $(1, 1)$ is $(3, 3)$.

6. Determine the slope of the tangent plane to the surface defined by $f(x, y) = x^2 + 3xy + y^2$ at the point $(1, 2)$.

Solution: The slope of the tangent plane is given by the gradient ∇f:

$$\nabla f = \left(\frac{\partial f}{\partial x}, \frac{\partial f}{\partial y} \right)$$

Compute:

$$\frac{\partial f}{\partial x} = 2x + 3y$$

$$\frac{\partial f}{\partial y} = 3x + 2y$$

Evaluate at $(1, 2)$:

$$\frac{\partial f}{\partial x}(1, 2) = 2 \cdot 1 + 3 \cdot 2 = 8$$

$$\frac{\partial f}{\partial y}(1, 2) = 3 \cdot 1 + 2 \cdot 2 = 7$$

Thus, the gradient at $(1, 2)$ is $(8, 7)$, indicating the steepest slopes of the tangent plane in the x and y directions.

Chapter 7

Gradient and Directional Derivatives

Practice Problems 1

1. Compute the gradient of the function:

$$f(x, y, z) = x^2 y + yz^3 + zx^2$$

2. Find the directional derivative of the function:

$$f(x, y) = \ln(x^2 + y^2)$$

at the point $(1, 1)$ in the direction of $\mathbf{v} = \langle 1, 0 \rangle$.

3. Given the function $g(x, y) = e^{xy}$, calculate the gradient at $(0, 2)$.

4. Consider the function $h(x, y, z) = xyz - x^3z + y^2$, find the gradient and evaluate it at the point $(1, 0, -1)$.

5. Determine the gradient descent update step for the function:
$$f(x, y) = 3x^2 + 2y^2 + 4xy$$
starting from point $(1, 1)$ with a learning rate $\alpha = 0.01$.

6. Compute the directional derivative of the function:
$$p(x, y, z) = x^2 + y^2 + z^2$$
at the point $(1, 2, 2)$ in the direction of $\mathbf{a} = \langle 1, 1, 1 \rangle$.

Answers 1

1. Compute the gradient of the function:
$$f(x, y, z) = x^2y + yz^3 + zx^2$$

Solution: To find the gradient ∇f, compute the partial derivatives with respect to x, y, and z.
$$\frac{\partial f}{\partial x} = 2xy + 2zx, \quad \frac{\partial f}{\partial y} = x^2 + z^3, \quad \frac{\partial f}{\partial z} = 3yz^2 + x^2$$

Therefore,
$$\nabla f = (2xy + 2zx, x^2 + z^3, 3yz^2 + x^2).$$

2. Find the directional derivative of the function:

$$f(x, y) = \ln(x^2 + y^2)$$

at the point $(1, 1)$ in the direction of $\mathbf{v} = \langle 1, 0 \rangle$.

Solution: First, compute the gradient $\nabla f = \left(\frac{2x}{x^2+y^2}, \frac{2y}{x^2+y^2} \right)$. Evaluating at $(1, 1)$:

$$\nabla f(1, 1) = \left(\frac{2 \cdot 1}{1^2 + 1^2}, \frac{2 \cdot 1}{1^2 + 1^2} \right) = \left(\frac{1}{1}, \frac{1}{1} \right) = (1, 1)$$

Normalize $\mathbf{v} = \langle 1, 0 \rangle$:

$$\mathbf{u} = \langle 1, 0 \rangle \text{ (already a unit vector)}$$

The directional derivative is:

$$D_{\mathbf{u}} f(1, 1) = (1, 1) \cdot \langle 1, 0 \rangle = 1 \cdot 1 + 1 \cdot 0 = 1$$

3. Given the function $g(x, y) = e^{xy}$, calculate the gradient at $(0, 2)$.
   Solution:

$$\frac{\partial g}{\partial x} = y e^{xy}, \quad \frac{\partial g}{\partial y} = x e^{xy}$$

Evaluating at $(0, 2)$:

$$\nabla g(0, 2) = (2 \cdot e^{0 \cdot 2}, 0 \cdot e^{0 \cdot 2}) = (2, 0)$$

4. Consider the function $h(x, y, z) = xyz - x^3 z + y^2$, find the gradient and evaluate it at the point $(1, 0, -1)$.
   Solution: Find partial derivatives:

$$\frac{\partial h}{\partial x} = yz - 3x^2 z, \quad \frac{\partial h}{\partial y} = xz + 2y, \quad \frac{\partial h}{\partial z} = xy - x^3$$

Evaluating at $(1, 0, -1)$:

$$\nabla h(1, 0, -1) = (0 - 3 \cdot 1^2 \cdot (-1), 1 \cdot (-1) + 0, 1 \cdot 0 - 1^3) = (3, -1, -1)$$

5. Determine the gradient descent update step for the function:

$$f(x, y) = 3x^2 + 2y^2 + 4xy$$

starting from point $(1, 1)$ with a learning rate $\alpha = 0.01$.
Solution: First, find gradient components:

$$\frac{\partial f}{\partial x} = 6x + 4y, \quad \frac{\partial f}{\partial y} = 4x + 4y$$

Evaluating at $(1, 1)$:

$$\nabla f(1, 1) = (6 \cdot 1 + 4 \cdot 1, 4 \cdot 1 + 4 \cdot 1) = (10, 8)$$

Update rule:

$$\mathbf{x}_{n+1} = \mathbf{x}_n - \alpha \nabla f$$

$$\mathbf{x}_{n+1} = (1, 1) - 0.01 \cdot (10, 8) = (0.9, 0.92)$$

6. Compute the directional derivative of the function:

$$p(x, y, z) = x^2 + y^2 + z^2$$

at the point $(1, 2, 2)$ in the direction of $\mathbf{a} = \langle 1, 1, 1 \rangle$.
Solution: compute the gradient:

$$\nabla p = (2x, 2y, 2z)$$

Evaluate at $(1, 2, 2)$:
$$\nabla p(1, 2, 2) = (2 \cdot 1, 2 \cdot 2, 2 \cdot 2) = (2, 4, 4)$$

Normalize $\mathbf{a} = \langle 1, 1, 1 \rangle$:
$$\text{Magnitude of } \mathbf{a} = \sqrt{1^2 + 1^2 + 1^2} = \sqrt{3}$$
$$\mathbf{u} = \left(\frac{1}{\sqrt{3}}, \frac{1}{\sqrt{3}}, \frac{1}{\sqrt{3}} \right)$$

Directional derivative is:
$$D_{\mathbf{u}} p = (2, 4, 4) \cdot \left(\frac{1}{\sqrt{3}}, \frac{1}{\sqrt{3}}, \frac{1}{\sqrt{3}} \right)$$
$$= \frac{2}{\sqrt{3}} + \frac{4}{\sqrt{3}} + \frac{4}{\sqrt{3}} = \frac{10}{\sqrt{3}}$$

Practice Problems 2

1. Compute the gradient of the following function:
$$f(x, y, z) = 4x^2 + y^2 + z^2$$

2. Find the directional derivative of the function
$$g(x, y) = x^3 y - y^2$$
at the point $(2, 1)$ in the direction of the vector $\mathbf{v} = \langle 1, 2 \rangle$.

3. For the function $h(x, y) = e^x \sin(y)$, calculate the gradient and evaluate it at the point $(0, \pi/2)$.

4. Determine the gradient vector of the function

$$i(x, y, z) = x \ln(y) + z^3$$

5. Compute the gradient and determine the direction of the steepest descent for

$$j(x, y) = x^4 - 2xy + y^2$$

when evaluated at the point (1, -1).

6. Using the function $k(x, y) = x^2 + xy + y^2$, find the directional derivative at the point (1, 1) in the direction of $\mathbf{w} = \langle -1, 1 \rangle$.

Answers 2

1. Compute the gradient of the following function:

$$f(x, y, z) = 4x^2 + y^2 + z^2$$

Solution: The gradient ∇f is given by:

$$\nabla f = \left(\frac{\partial f}{\partial x}, \frac{\partial f}{\partial y}, \frac{\partial f}{\partial z} \right)$$

$$= \left(\frac{\partial}{\partial x}(4x^2), \frac{\partial}{\partial y}(y^2), \frac{\partial}{\partial z}(z^2) \right)$$

$$= (8x, 2y, 2z)$$

Therefore,

$$\nabla f = (8x, 2y, 2z).$$

2. Find the directional derivative of the function

$$g(x, y) = x^3 y - y^2$$

at the point $(2, 1)$ in the direction of the vector $\mathbf{v} = \langle 1, 2 \rangle$.
Solution: First, compute the gradient ∇g:

$$\nabla g = \left(\frac{\partial g}{\partial x}, \frac{\partial g}{\partial y} \right)$$

$$= \left(3x^2 y, x^3 - 2y \right)$$

Evaluate at the point $(2, 1)$:

$$\nabla g(2, 1) = (3 \cdot 2^2 \cdot 1, 2^3 - 2 \cdot 1) = (12, 6)$$

Normalize $\mathbf{v} = \langle 1, 2 \rangle$:

$$\mathbf{u} = \frac{\langle 1, 2 \rangle}{\sqrt{1^2 + 2^2}} = \left\langle \frac{1}{\sqrt{5}}, \frac{2}{\sqrt{5}} \right\rangle$$

Compute the directional derivative:

$$D_{\mathbf{u}} g(2, 1) = \nabla g(2, 1) \cdot \mathbf{u}$$

$$= (12, 6) \cdot \left\langle \frac{1}{\sqrt{5}}, \frac{2}{\sqrt{5}} \right\rangle$$

$$= \frac{12}{\sqrt{5}} + \frac{12}{\sqrt{5}} = \frac{24}{\sqrt{5}}$$

Therefore, the directional derivative is $\frac{24}{\sqrt{5}}$.

3. For the function $h(x, y) = e^x \sin(y)$, calculate the gradient and evaluate it at the point $(0, \pi/2)$.
Solution: Compute the gradient:

$$\nabla h = \left(\frac{\partial h}{\partial x}, \frac{\partial h}{\partial y} \right)$$

$$= (e^x \sin(y), e^x \cos(y))$$

Evaluate at the point $(0, \pi/2)$:

$$\nabla h(0, \pi/2) = (e^0 \sin(\pi/2), e^0 \cos(\pi/2)) = (1 \cdot 1, 1 \cdot 0) = (1, 0)$$

Therefore, the gradient at $(0, \pi/2)$ is $(1, 0)$.

4. Determine the gradient vector of the function

$$i(x, y, z) = x \ln(y) + z^3$$

Solution: Compute the gradient:

$$\nabla i = \left(\frac{\partial i}{\partial x}, \frac{\partial i}{\partial y}, \frac{\partial i}{\partial z} \right)$$

$$= \left(\ln(y), \frac{x}{y}, 3z^2 \right)$$

Therefore, the gradient vector is $\left(\ln(y), \frac{x}{y}, 3z^2 \right)$.

5. Compute the gradient and determine the direction of the steepest descent for

$$j(x, y) = x^4 - 2xy + y^2$$

when evaluated at the point (1, -1).
Solution: Compute the gradient:

$$\nabla j = \left(\frac{\partial j}{\partial x}, \frac{\partial j}{\partial y} \right)$$

$$= \left(4x^3 - 2y, -2x + 2y \right)$$

Evaluate at (1, -1):

$$\nabla j(1, -1) = (4 \cdot 1^3 - 2 \cdot (-1), -2 \cdot 1 + 2 \cdot (-1))$$

$$= (4 + 2, -2 - 2) = (6, -4)$$

The direction of steepest descent is $-\nabla j(1, -1) = (-6, 4)$.

6. Using the function $k(x, y) = x^2 + xy + y^2$, find the directional derivative at the point (1, 1) in the direction of $\mathbf{w} = \langle -1, 1 \rangle$.
Solution: Compute the gradient:

$$\nabla k = \left(\frac{\partial k}{\partial x}, \frac{\partial k}{\partial y} \right)$$

$$= (2x + y, x + 2y)$$

Evaluate at (1, 1):

$$\nabla k(1, 1) = (2 \cdot 1 + 1, 1 + 2 \cdot 1) = (3, 3)$$

Normalize $\mathbf{w} = \langle -1, 1 \rangle$:

$$\mathbf{u} = \frac{\langle -1, 1 \rangle}{\sqrt{(-1)^2 + 1^2}} = \left\langle -\frac{1}{\sqrt{2}}, \frac{1}{\sqrt{2}} \right\rangle$$

Compute the directional derivative:

$$D_{\mathbf{u}} k(1, 1) = \nabla k(1, 1) \cdot \mathbf{u}$$

$$= (3, 3) \cdot \left\langle -\frac{1}{\sqrt{2}}, \frac{1}{\sqrt{2}} \right\rangle$$

$$= 3 \cdot -\frac{1}{\sqrt{2}} + 3 \cdot \frac{1}{\sqrt{2}} = 0$$

Therefore, the directional derivative is 0.

Practice Problems 3

1. Compute the gradient of the function:

$$f(x, y) = x^3 y + 2xy^2 - x + y$$

2. Calculate the directional derivative of the function $f(x, y) = x^2y + y^3$ at the point $(1, 2)$ in the direction of the vector $\mathbf{v} = \langle 1, 1 \rangle$.

3. Determine if the point $(2, -1)$ is a critical point of the function $f(x, y) = x^2 + y^2 - 2x - 4y$.

4. For the function $f(x, y, z) = xy + yz + zx$, find the gradient and then evaluate its magnitude at the point $(1, -1, 1)$.

5. Consider the function $g(x, y) = 4x^2 + xy + y^2$. Compute the gradient and determine the direction of the steepest increase at the point $(2, 3)$.

6. Use the gradient descent update rule to find the next point if starting from $\mathbf{x}_0 = (3, 1)$ for the function $h(x, y) = x^2 + y^2$ with a learning rate of $\alpha = 0.1$.

Answers 3

1. Compute the gradient of the function:

$$f(x, y) = x^3 y + 2xy^2 - x + y$$

   Solution: First, find the partial derivatives:

$$\frac{\partial f}{\partial x} = 3x^2 y + 2y^2 - 1$$

$$\frac{\partial f}{\partial y} = x^3 + 4xy + 1$$

   Therefore, the gradient is:

$$\nabla f = \left(3x^2 y + 2y^2 - 1, x^3 + 4xy + 1\right)$$

2. Calculate the directional derivative of $f(x, y) = x^2 y + y^3$ at $(1, 2)$ in the direction of $\mathbf{v} = \langle 1, 1 \rangle$.

   Solution: Normalize $\mathbf{v}$:

$$\mathbf{u} = \frac{1}{\sqrt{1^2 + 1^2}} \langle 1, 1 \rangle = \left\langle \frac{\sqrt{2}}{2}, \frac{\sqrt{2}}{2} \right\rangle$$

   Compute the gradient:

$$\nabla f = \left(2xy, x^2 + 3y^2\right)$$

   Evaluate at $(1, 2)$:

$$\nabla f(1, 2) = (4, 13)$$

   The directional derivative is:

$$D_{\mathbf{u}} f(1, 2) = \nabla f(1, 2) \cdot \mathbf{u} = 4 \times \frac{\sqrt{2}}{2} + 13 \times \frac{\sqrt{2}}{2}$$

$$= \frac{34}{\sqrt{2}} = 17\sqrt{2}$$

3. Determine if $(2, -1)$ is a critical point of $f(x, y) = x^2 + y^2 - 2x - 4y$.

   Solution: Compute the gradient:

$$\nabla f = (2x - 2, 2y - 4)$$

   Evaluate at $(2, -1)$:

$$\nabla f(2, -1) = (4 - 2, -2 - 4) = (2, -6)$$

   Since $\nabla f(2, -1) \neq (0, 0)$, $(2, -1)$ is not a critical point.

4. For $f(x, y, z) = xy + yz + zx$, find the gradient and evaluate its magnitude at $(1, -1, 1)$.

   Solution: Compute the gradient:

$$\nabla f = (y + z, x + z, x + y)$$

   Evaluate at $(1, -1, 1)$:

$$\nabla f(1, -1, 1) = (0, 2, 0)$$

   Magnitude is:

$$\|\nabla f(1, -1, 1)\| = \sqrt{0^2 + 2^2 + 0^2} = 2$$

5. For $g(x, y) = 4x^2 + xy + y^2$, compute the gradient and determine the direction of steepest increase at $(2, 3)$.

   Solution: Compute the gradient:
   $$\nabla g = (8x + y, x + 2y)$$

   Evaluate at $(2, 3)$:
   $$\nabla g(2, 3) = (19, 8)$$

   The direction of steepest increase is $\nabla g(2, 3) = (19, 8)$.

6. Use the gradient descent update rule to find the next point starting from $\mathbf{x}_0 = (3, 1)$ for $h(x, y) = x^2 + y^2$ with $\alpha = 0.1$.

   Solution: Compute the gradient:
   $$\nabla h = (2x, 2y)$$

   Evaluate at $(3, 1)$:
   $$\nabla h(3, 1) = (6, 2)$$

   Apply the update rule:
   $$\mathbf{x}_1 = (3, 1) - 0.1 \cdot (6, 2) = (3, 1) - (0.6, 0.2) = (2.4, 0.8)$$

   Thus, the next point is $\mathbf{x}_1 = (2.4, 0.8)$.

Chapter 8

Optimization Techniques

Practice Problems 1

1. Explain the convergence criteria for the gradient descent algorithm and derive the stopping condition in terms of the gradient norm.

2. Analyze the computational cost of Newton's method by discussing the implications of computing and inverting the Hessian matrix in high-dimensional spaces.

3. Given a function $f(x, y) = x^2 + y^2$, apply the gradient descent method to find the minimum starting from the point $(1, 1)$ with a learning rate $\alpha = 0.1$. Perform two iterations.

4. Derive the BFGS update formula for the inverse Hessian approximation and discuss how it improves upon the standard gradient descent.

5. Compare and contrast the convergence properties of stochastic gradient descent (SGD) with batch gradient descent, particularly in the context of large datasets.

6. Using the function $f(x) = x^4 - 3x^3 + 2$, apply Newton's method to find a critical point starting with $x_0 = 1$. Perform one update step.

Answers 1

1. Explain the convergence criteria for the gradient descent algorithm and derive the stopping condition in terms of the gradient norm.

   Solution:
   The convergence criteria for gradient descent is determined by the magnitude of the gradient. The algorithm stops when the gradient's norm is sufficiently small, indicating that the parameters are close to a local minimum. Formally, if $\nabla f(x^{(t)})$ is the gradient at iteration t, we stop when:

   $$\|\nabla f(x^{(t)})\| \leq \epsilon$$

   where ϵ is a predefined small threshold.

2. Analyze the computational cost of Newton's method by discussing the implications of computing and inverting the Hessian matrix in high-dimensional spaces.

Solution:
Newton's method requires computing the Hessian matrix $\nabla^2 f(x)$ and its inverse at each iteration. In an n-dimensional space, the Hessian is an $n \times n$ matrix. Computing the Hessian costs $\mathcal{O}(n^2)$ operations, and inverting it costs $\mathcal{O}(n^3)$. In high-dimensional problems, this can be computationally prohibitive, thus limiting the method's practicality.

3. Given a function $f(x, y) = x^2 + y^2$, apply the gradient descent method to find the minimum starting from the point $(1, 1)$ with a learning rate $\alpha = 0.1$. Perform two iterations.

Solution:
The gradient is $\nabla f(x, y) = (2x, 2y)$. Starting at $(x_0, y_0) = (1, 1)$, the updates are:

$$(x_1, y_1) = (x_0, y_0) - \alpha \nabla f(x_0, y_0) = (1, 1) - 0.1 \cdot (2, 2) = (0.8, 0.8)$$

For the second iteration:

$$(x_2, y_2) = (x_1, y_1) - \alpha \nabla f(x_1, y_1) = (0.8, 0.8) - 0.1 \cdot (1.6, 1.6) = (0.64, 0.64)$$

4. Derive the BFGS update formula for the inverse Hessian approximation and discuss how it improves upon the standard gradient descent.

Solution:
The BFGS update formula for the inverse Hessian approximation $B^{(t+1)}$ is:

$$B^{(t+1)} = B^{(t)} + \frac{y^{(t)} y^{(t)\top}}{y^{(t)\top} s^{(t)}} - \frac{B^{(t)} s^{(t)} s^{(t)\top} B^{(t)}}{s^{(t)\top} B^{(t)} s^{(t)}}$$

where $s^{(t)} = x^{(t+1)} - x^{(t)}$ and $y^{(t)} = \nabla f(x^{(t+1)}) - \nabla f(x^{(t)})$. The BFGS method approximates the inverse Hessian, offering faster convergence than standard gradient descent by effectively adjusting the step size.

5. Compare and contrast the convergence properties of stochastic gradient descent (SGD) with batch gradient descent, particularly in the context of large datasets.

Solution:
SGD approximates the gradient using a minibatch, providing stochastic updates that introduce variability. This can help escape local minima and improve convergence speed, particularly in large datasets. In contrast, batch gradient descent uses the entire dataset for each update, ensuring stable and deterministic convergence but often at the cost of slower updates, especially when datasets are large.

6. Using the function $f(x) = x^4 - 3x^3 + 2$, apply Newton's method to find a critical point starting with $x_0 = 1$. Perform one update step.

Solution:
The derivative is $f'(x) = 4x^3 - 9x^2$ and the second derivative is $f''(x) = 12x^2 - 18x$. Start with $x_0 = 1$:

$$x_1 = x_0 - \frac{f'(x_0)}{f''(x_0)} = 1 - \frac{4(1)^3 - 9(1)^2}{12(1)^2 - 18(1)}$$

$$= 1 - \frac{-5}{-6} = 1 + \frac{5}{6} = \frac{11}{6}$$

Practice Problems 2

1. Consider the function $f(x, y) = x^2 y - y^3$. Compute the gradient ∇f.

2. Given the function $g(x) = \frac{1}{2}x^4 - 3x^2 + 5$, perform one iteration of Newton's method starting at $x_0 = 1$.

3. For the logistic loss function $L(w) = \log(1 + e^{-w^\top x})$, compute the gradient with respect to w.

4. Determine the update rule for the BFGS algorithm given $B^{(t)} = I$, where I is the identity matrix, in the initial step.

5. Derive the convergence criterion formula for the gradient descent method and explain its significance.

6. Explain how stochastic gradient descent (SGD) is advantageous for mini-batch training compared to traditional gradient descent.

Answers 2

1. Consider the function $f(x, y) = x^2 y - y^3$. Compute the gradient ∇f.
   Solution:

   $$\nabla f = \left(\frac{\partial f}{\partial x}, \frac{\partial f}{\partial y} \right)$$

   $$\frac{\partial f}{\partial x} = \frac{\partial}{\partial x}(x^2 y - y^3) = 2xy$$

   $$\frac{\partial f}{\partial y} = \frac{\partial}{\partial y}(x^2 y - y^3) = x^2 - 3y^2$$

   Therefore,

   $$\nabla f = (2xy, x^2 - 3y^2).$$

2. Given the function $g(x) = \frac{1}{2}x^4 - 3x^2 + 5$, perform one iteration of Newton's method starting at $x_0 = 1$.
   Solution:

   $$g'(x) = 2x^3 - 6x$$

   $$g''(x) = 6x^2 - 6$$

   Using Newton's method:

   $$x_1 = x_0 - \frac{g'(x_0)}{g''(x_0)} = 1 - \frac{2(1)^3 - 6(1)}{6(1)^2 - 6}$$

   $$= 1 - \frac{2 - 6}{6 - 6} = 1$$

   (Note: The division by zero suggests a deeper problem, indicating the point $x_0 = 1$ is not suitable for this iteration as it's an inflection point based on $g''(x) = 0$)

3. For the logistic loss function $L(w) = \log(1 + e^{-w^\top x})$, compute the gradient with respect to w.
   Solution:

   $$\frac{\partial L}{\partial w} = -\frac{e^{-w^\top x}}{1 + e^{-w^\top x}} \cdot (-x)$$

   $$= \frac{1}{1 + e^{-w^\top x}} x$$

   Therefore, the gradient is:

   $$\frac{\partial L}{\partial w} = \sigma(w^\top x)x$$

   where $\sigma(z) = \frac{1}{1+e^{-z}}$.

4. Determine the update rule for the BFGS algorithm given $B^{(t)} = I$, where I is the identity matrix, in the initial step.
   Solution:

   $$B^{(t+1)} = B^{(t)} + \frac{y^{(t)} y^{(t)\top}}{y^{(t)\top} s^{(t)}} - \frac{B^{(t)} s^{(t)} s^{(t)\top} B^{(t)}}{s^{(t)\top} B^{(t)} s^{(t)}}$$

   Initially, $B^{(t)} = I$:

   $$B^{(t+1)} = I + \frac{y^{(t)} y^{(t)\top}}{y^{(t)\top} s^{(t)}} - \frac{s^{(t)} s^{(t)\top}}{s^{(t)\top} s^{(t)}}$$

5. Derive the convergence criterion formula for the gradient descent method and explain its significance.
   Solution: The convergence criterion for gradient descent is given by:

   $$\|\nabla f(x^{(t)})\| \leq \epsilon$$

   This criterion ensures that the magnitude of the gradient is sufficiently small, indicating proximity to a local minimum. It is crucial because it prevents unnecessary iterations once an optimal or satisfactorily close point is reached, balancing computational efficiency and precision.

6. Explain how stochastic gradient descent (SGD) is advantageous for mini-batch training compared to traditional gradient descent.
   Solution: Stochastic Gradient Descent (SGD) differs from traditional gradient descent in using a randomly selected mini-batch of data for each update rather than the entire dataset. This allows:
   - Faster computation per iteration because only a subset of data is used,
   - Greater efficiency in handling large datasets due to reduced memory usage,
   - Introduction of noise in the updates, which can help achieving better generalization by escaping from shallow local minima. These advantages make SGD especially beneficial in real-time or online learning scenarios.

Practice Problems 3

1. Derive the update rule for the gradient descent algorithm using the gradient of a scalar function $f(x)$.

2. Prove the convergence criteria for gradient descent. Show how the magnitude of the gradient helps in determining convergence.

3. Explain why Newton's method requires the Hessian matrix to be invertible. What happens if the Hessian is singular?

4. Derive the update formula for the BFGS algorithm used in quasi-Newton methods.

5. Discuss the advantages of using stochastic gradient descent (SGD) over batch gradient descent. Include consideration of computational cost and data size.

6. Demonstrate how momentum can be incorporated into the gradient descent update rule. Explain the effect of momentum on convergence.

Answers 3

1. **Derive the update rule for the gradient descent algorithm using the gradient of a scalar function $f(x)$.**

   Solution: The gradient descent algorithm aims to find the minimum of a function $f(x)$. The basic update rule is given by moving opposite to the direction of the gradient:

   $$x^{(t+1)} = x^{(t)} - \alpha \nabla f(x^{(t)})$$

   Here, $\nabla f(x^{(t)})$ is the gradient of f at $x^{(t)}$, and α is the learning rate. It ensures that with each update, the parameter moves in the direction of the steepest descent to find the local or global minimum.

2. **Prove the convergence criteria for gradient descent. Show how the magnitude of the gradient helps in determining convergence.**

   Solution: Convergence in gradient descent is determined when changes in the parameter become negligibly small. The criterion is:

   $$\|\nabla f(x^{(t)})\| \leq \epsilon$$

   This implies that when the magnitude of $\nabla f(x)$ becomes less than or equal to a small threshold ϵ, the algorithm assumes convergence. This is because the gradient vector close to zero suggests that the function is approaching a local minimum.

3. **Explain why Newton's method requires the Hessian matrix to be invertible. What happens if the Hessian is singular?**

   Solution: Newton's method utilizes the Hessian matrix $\nabla^2 f(x)$ for its update rule:

   $$x^{(t+1)} = x^{(t)} - \left(\nabla^2 f(x^{(t)})\right)^{-1} \nabla f(x^{(t)})$$

   The inverse of the Hessian $\left(\nabla^2 f(x)\right)^{-1}$ is critical to compute direction and step length. If the Hessian is singular (non-invertible), the method fails as it relies on matrix inversion. In such a case, quasi-Newton methods or regularization techniques are employed to approximate or adjust the Hessian.

4. **Derive the update formula for the BFGS algorithm used in quasi-Newton methods.**

   Solution: The BFGS algorithm updates the inverse Hessian approximation $B^{(t)}$ using:

   $$B^{(t+1)} = B^{(t)} + \frac{y^{(t)}y^{(t)\top}}{y^{(t)\top}s^{(t)}} - \frac{B^{(t)}s^{(t)}s^{(t)\top}B^{(t)}}{s^{(t)\top}B^{(t)}s^{(t)}}$$

   Where:
   $$s^{(t)} = x^{(t+1)} - x^{(t)}, \quad y^{(t)} = \nabla f(x^{(t+1)}) - \nabla f(x^{(t)})$$

   This maintains a balance between efficiency and convergence by updating the approximation info about the curvature of f.

5. **Discuss the advantages of using stochastic gradient descent (SGD) over batch gradient descent. Include consideration of computational cost and data size.**

   Solution: Stochastic Gradient Descent (SGD) uses a random subset (or minibatch) of the data for each update, unlike Batch Gradient Descent which uses the entire dataset. This results in:
   - Lower computational cost per update since only a subset is used.
   - Faster convergence in practice since it introduces noise, helping escape local minima.
   - Scalability to very large datasets where using the entire dataset is impractical.

6. **Demonstrate how momentum can be incorporated into the gradient descent update rule. Explain the effect of momentum on convergence.**

   Solution: Momentum modifies the update rule by considering past gradients:

   $$v^{(t)} = \beta v^{(t-1)} + \nabla f(x^{(t)})$$

   $$x^{(t+1)} = x^{(t)} - \alpha v^{(t)}$$

   Where v is velocity, β is the momentum factor, usually between 0.7 and 0.9. Momentum makes updates go faster in consistent gradient directions while damping oscillations, hence improving convergence speed and stability.

Chapter 9

Second Derivatives and the Hessian

Practice Problems 1

1. Given a univariate function $f(x) = 3x^4 - 4x^3 + x - 5$, find the second derivative $f''(x)$.

2. Consider the multivariate function $f(x, y) = xy^2 + 3x^2y + 4$. Compute the Hessian matrix $\nabla^2 f(x, y)$.

3. For the function $f(x) = e^{x^2}$, find the second derivative $f''(x)$ using the chain rule.

4. Given the function $f(x, y) = 2x^3 + 3xy + y^3$, determine whether this point $(0,0)$ is a saddle point.

5. Calculate the eigenvalues of the Hessian matrix for $f(x,y) = x^2 + xy + y^2$ and determine the curvature at any point.

6. Use Newton's Method to find an approximation of a root for $f'(x) = 3x^2 - 12x$ starting with $x_0 = 2$.

Answers 1

1. **Find $f''(x)$ for $f(x) = 3x^4 - 4x^3 + x - 5$:**

$$f'(x) = \frac{d}{dx}(3x^4) - \frac{d}{dx}(4x^3) + \frac{d}{dx}(x) - \frac{d}{dx}(5)$$

$$= 12x^3 - 12x^2 + 1$$

Now find $f''(x)$:

$$f''(x) = \frac{d}{dx}(12x^3) - \frac{d}{dx}(12x^2) + \frac{d}{dx}(1)$$

$$= 36x^2 - 24x$$

Therefore, $f''(x) = 36x^2 - 24x$.

2. **Hessian matrix for $f(x,y) = xy^2 + 3x^2y + 4$:** Calculate the second-order partial derivatives:

$$f''_{xx} = \frac{\partial^2 f}{\partial x^2} = \frac{\partial}{\partial x}(y^2 + 6xy) = 6y$$

$$f''_{yy} = \frac{\partial^2 f}{\partial y^2} = \frac{\partial}{\partial y}(2xy + 3x^2) = 2x$$

$$f''_{xy} = \frac{\partial^2 f}{\partial x \partial y} = \frac{\partial}{\partial y}(2xy + 3x^2) = 2x$$

$$f''_{yx} = \frac{\partial^2 f}{\partial y \partial x} = \frac{\partial}{\partial x}(y^2 + 6xy) = 2y$$

Hessian matrix:

$$\nabla^2 f(x,y) = \begin{bmatrix} 6y & 2x \\ 2x & 2y \end{bmatrix}$$

84

3. **Find $f''(x)$ for $f(x) = e^{x^2}$ using chain rule:**

$$f'(x) = \frac{d}{dx}(e^{x^2}) = e^{x^2} \cdot \frac{d}{dx}(x^2) = e^{x^2} \cdot 2x$$

$$f''(x) = \frac{d}{dx}(2xe^{x^2})$$

Use product rule:

$$f''(x) = 2 \cdot e^{x^2} + 2x \cdot e^{x^2} \cdot 2x = 2e^{x^2} + 4x^2 e^{x^2}$$

Therefore, $f''(x) = 2e^{x^2} + 4x^2 e^{x^2}$.

4. **Determine if $(0,0)$ is a saddle point of $f(x,y) = 2x^3 + 3xy + y^3$:** First, find the Hessian at $(0,0)$:

$$f''_{xx} = \frac{\partial^2 f}{\partial x^2} = 6x \Rightarrow f_{xx}(0,0) = 0$$

$$f''_{yy} = \frac{\partial^2 f}{\partial y^2} = 6y \Rightarrow f_{yy}(0,0) = 0$$

$$f''_{xy} = \frac{\partial^2 f}{\partial x \partial y} = 3 \Rightarrow f_{xy}(0,0) = 3$$

Hessian at $(0,0)$:

$$\nabla^2 f(0,0) = \begin{bmatrix} 0 & 3 \\ 3 & 0 \end{bmatrix}$$

Since the determinant $|H| = 0 \cdot 0 - 3 \cdot 3 = -9$, it is negative, indicating a saddle point.

5. **Eigenvalues of the Hessian for $f(x,y) = x^2 + xy + y^2$:** Calculate the Hessian matrix:

$$f''_{xx} = 2, \quad f''_{yy} = 2, \quad f''_{xy} = 1$$

$$\nabla^2 f(x,y) = \begin{bmatrix} 2 & 1 \\ 1 & 2 \end{bmatrix}$$

Find the eigenvalues λ from $\det(\nabla^2 f - \lambda I) = 0$:

$$\det\left(\begin{bmatrix} 2-\lambda & 1 \\ 1 & 2-\lambda \end{bmatrix}\right) = (2-\lambda)^2 - 1 = 0$$

$$(2-\lambda)^2 - 1 = \lambda^2 - 4\lambda + 3 = 0 \Rightarrow (\lambda - 3)(\lambda - 1) = 0$$

Eigenvalues are $\lambda_1 = 3$ and $\lambda_2 = 1$, both positive, indicating positive definite Hessian, thus the function is convex.

6. **Newton's Method for $f'(x) = 3x^2 - 12x$ with $x_0 = 2$:** Solve for roots of $f'(x) = 3x(x-4)$:

$$x^{(t+1)} = x^{(t)} - \frac{f'(x^{(t)})}{f''(x^{(t)})}$$

$$f''(x) = \frac{d}{dx}(3x^2 - 12x) = 6x - 12$$

Using $x_0 = 2$:

$$f'(x_0) = 3(2)^2 - 12(2) = 12 - 24 = -12$$

$$f''(x_0) = 6(2) - 12 = 0$$

As $f''(x_0) = 0$, the method cannot proceed, suggesting to try another x_0.

Practice Problems 2

1. Compute the second derivative of the following univariate function and determine the intervals of convexity and concavity:
$$f(x) = x^3 - 3x^2 + 4x - 2$$

2. For the function $g(x, y) = 3x^2y + 2y^3 - 6xy$, compute the Hessian matrix and determine if the function is locally convex at the point $(1, 1)$.

3. Determine the critical points of $h(x, y) = x^2 + xy + y^2$ and use the Hessian to classify each as a local minimum, local maximum, or saddle point.

4. Using eigenvalue decomposition, express the Hessian of $f(x, y) = 2x^2 + 4xy + y^2$ and determine the nature of its critical points.

5. For the function $f(x, y) = e^{x^2 + y^2}$, compute the Hessian matrix and analyze the curvature at the origin.

6. Apply Newton's method to find one iteration step for the function $f(x) = x^2 - 4x + 3$ starting at $x^{(0)} = 0$.

Answers 2

1. Compute the second derivative of the following univariate function and determine the intervals of convexity and concavity:

$$f(x) = x^3 - 3x^2 + 4x - 2$$

Solution: First, compute the first derivative:

$$f'(x) = \frac{d}{dx}(x^3) - \frac{d}{dx}(3x^2) + \frac{d}{dx}(4x) - \frac{d}{dx}(2)$$

$$= 3x^2 - 6x + 4$$

Now, compute the second derivative:

$$f''(x) = \frac{d}{dx}(3x^2) - \frac{d}{dx}(6x)$$

$$= 6x - 6$$

Set $f''(x) = 0$ to find inflection points:

$$6x - 6 = 0 \Rightarrow x = 1$$

Test intervals around $x = 1$:
- For $x < 1$, pick $x = 0$: $f''(0) = 6(0) - 6 = -6$ (concave)
- For $x > 1$, pick $x = 2$: $f''(2) = 6(2) - 6 = 6$ (convex)

Therefore, $f(x)$ is concave on $(-\infty, 1)$ and convex on $(1, \infty)$.

2. For the function $g(x, y) = 3x^2y + 2y^3 - 6xy$, compute the Hessian matrix and determine if the function is locally convex at the point $(1, 1)$.

   Solution: First, compute the first-order partial derivatives:

   $$g_x = \frac{\partial}{\partial x}(3x^2y) + \frac{\partial}{\partial x}(2y^3) + \frac{\partial}{\partial x}(-6xy) = 6xy - 6y$$

   $$g_y = \frac{\partial}{\partial y}(3x^2y) + \frac{\partial}{\partial y}(2y^3) + \frac{\partial}{\partial y}(-6xy) = 3x^2 + 6y^2 - 6x$$

   Calculate the second-order partial derivatives:

   $$g_{xx} = \frac{\partial}{\partial x}(6xy - 6y) = 6y$$

   $$g_{yy} = \frac{\partial}{\partial y}(3x^2 + 6y^2 - 6x) = 12y$$

   $$g_{xy} = \frac{\partial}{\partial y}(6xy - 6y) = 6x - 6$$

   The Hessian matrix is:
   $$\nabla^2 g(x, y) = \begin{bmatrix} g_{xx} & g_{xy} \\ g_{yx} & g_{yy} \end{bmatrix} = \begin{bmatrix} 6y & 6x - 6 \\ 6x - 6 & 12y \end{bmatrix}$$

   Evaluate at $(1, 1)$:
   $$\nabla^2 g(1, 1) = \begin{bmatrix} 6 & 0 \\ 0 & 12 \end{bmatrix}$$

   The eigenvalues are 6 and 12, both positive, indicating local convexity at $(1, 1)$.

3. Determine the critical points of $h(x, y) = x^2 + xy + y^2$ and use the Hessian to classify each as a local minimum, local maximum, or saddle point.

   Solution: First, find critical points by setting the first-order partial derivatives to zero:

   $$h_x = \frac{\partial}{\partial x}(x^2 + xy + y^2) = 2x + y = 0$$

   $$h_y = \frac{\partial}{\partial y}(x^2 + xy + y^2) = x + 2y = 0$$

   Solve the system:
   $$2x + y = 0 \quad (1)$$
   $$x + 2y = 0 \quad (2)$$

   From equation (1), $y = -2x$. Substitute into (2):

   $$x + 2(-2x) = 0 \Rightarrow x - 4x = 0 \Rightarrow -3x = 0 \Rightarrow x = 0$$

   $$y = -2(0) = 0$$

   Thus, the critical point is $(0, 0)$.

   Compute second-order partial derivatives for the Hessian:

   $$h_{xx} = \frac{\partial}{\partial x}(2x + y) = 2$$

   $$h_{yy} = \frac{\partial}{\partial y}(x + 2y) = 2$$

   88

$$h_{xy} = \frac{\partial}{\partial y}(2x + y) = 1$$

Hessian matrix:

$$\nabla^2 h(x, y) = \begin{bmatrix} 2 & 1 \\ 1 & 2 \end{bmatrix}$$

Calculate the determinant:

$$\det(\nabla^2 h) = (2)(2) - (1)(1) = 4 - 1 = 3 > 0$$

And since $h_{xx} > 0$, the point $(0, 0)$ is a local minimum.

4. Using eigenvalue decomposition, express the Hessian of $f(x, y) = 2x^2 + 4xy + y^2$ and determine the nature of its critical points.

Solution: Calculate the first-order partial derivatives:

$$f_x = \frac{\partial}{\partial x}(2x^2 + 4xy + y^2) = 4x + 4y$$

$$f_y = \frac{\partial}{\partial y}(2x^2 + 4xy + y^2) = 4x + 2y$$

Find the critical points by setting these to zero:

$$4x + 4y = 0 \quad (1)$$

$$4x + 2y = 0 \quad (2)$$

Subtract (2) from (1):

$$(4x + 4y) - (4x + 2y) = 0$$

$$2y = 0 \Rightarrow y = 0$$

$$4x + 0 = 0 \Rightarrow x = 0$$

Therefore, the only critical point is $(0, 0)$.

Hessian matrix:

$$f_{xx} = \frac{\partial}{\partial x}(4x + 4y) = 4$$

$$f_{yy} = \frac{\partial}{\partial y}(4x + 2y) = 2$$

$$f_{xy} = \frac{\partial}{\partial y}(4x + 4y) = 4$$

The Hessian matrix is:

$$\nabla^2 f(x, y) = \begin{bmatrix} 4 & 4 \\ 4 & 2 \end{bmatrix}$$

Perform eigenvalue decomposition:

$$\det(\nabla^2 f - \lambda I) = \begin{vmatrix} 4 - \lambda & 4 \\ 4 & 2 - \lambda \end{vmatrix} = 0$$

$$(4 - \lambda)(2 - \lambda) - 16 = 0$$

$$\lambda^2 - 6\lambda + 8 - 16 = 0$$

$$\lambda^2 - 6\lambda - 8 = 0$$

Solve the quadratic:

$$\lambda = \frac{6 \pm \sqrt{36 + 32}}{2} = \frac{6 \pm \sqrt{68}}{2}$$

Simplify:

$$\lambda_1, \lambda_2 = \frac{6 \pm 2\sqrt{17}}{2}$$

The eigenvalues indicate a saddle point since they have different signs.

5. For the function $f(x, y) = e^{x^2 + y^2}$, compute the Hessian matrix and analyze the curvature at the origin.

   Solution: Compute first derivatives:

$$f_x = \frac{\partial}{\partial x}(e^{x^2 + y^2}) = e^{x^2 + y^2} \cdot 2x$$

$$f_y = \frac{\partial}{\partial y}(e^{x^2 + y^2}) = e^{x^2 + y^2} \cdot 2y$$

Compute second-order derivatives:

$$f_{xx} = \frac{\partial}{\partial x}(e^{x^2 + y^2} \cdot 2x) = (e^{x^2 + y^2} \cdot 2x \cdot 2x) + (e^{x^2 + y^2} \cdot 2)$$

$$= 4x^2 e^{x^2 + y^2} + 2e^{x^2 + y^2}$$

$$f_{yy} = \frac{\partial}{\partial y}(e^{x^2 + y^2} \cdot 2y) = (e^{x^2 + y^2} \cdot 2y \cdot 2y) + (e^{x^2 + y^2} \cdot 2)$$

$$= 4y^2 e^{x^2 + y^2} + 2e^{x^2 + y^2}$$

$$f_{xy} = \frac{\partial}{\partial y}(e^{x^2 + y^2} \cdot 2x) = 4xy e^{x^2 + y^2}$$

The Hessian matrix is:

$$\nabla^2 f(x, y) = \begin{bmatrix} 4x^2 e^{x^2 + y^2} + 2e^{x^2 + y^2} & 4xy e^{x^2 + y^2} \\ 4xy e^{x^2 + y^2} & 4y^2 e^{x^2 + y^2} + 2e^{x^2 + y^2} \end{bmatrix}$$

Evaluate at the origin $(0, 0)$:

$$\nabla^2 f(0, 0) = \begin{bmatrix} 2 & 0 \\ 0 & 2 \end{bmatrix}$$

Both eigenvalues are positive, indicating that the function is locally convex at the origin.

6. Apply Newton's method to find one iteration step for the function $f(x) = x^2 - 4x + 3$ starting at $x^{(0)} = 0$.

   Solution: Compute the first and second derivative:

$$f'(x) = \frac{d}{dx}(x^2 - 4x + 3) = 2x - 4$$

$$f''(x) = \frac{d}{dx}(2x - 4) = 2$$

Newton's update for one iteration from $x^{(0)} = 0$:

$$x^{(1)} = x^{(0)} - \frac{f'(x^{(0)})}{f''(x^{(0)})}$$

$$= 0 - \frac{2(0) - 4}{2}$$

$$= 0 + \frac{4}{2}$$

$$= 2$$

Thus, after one iteration, the new estimate is $x^{(1)} = 2$.

Practice Problems 3

1. Given a function $f(x, y) = 3x^2y + xy^2 + y^3$, compute the Hessian matrix:

$$\nabla^2 f(x, y)$$

2. Determine if the function $g(x, y) = x^2 + 4xy + 4y^2$ is convex by analyzing its Hessian matrix.

3. For the function $h(x, y) = x^3 + 3xy + y^3$, find the critical points and classify them using the Hessian matrix.

4. Consider the function $u(x, y, z) = x^2y + yz^2 + xz$. Calculate the second-order partial derivatives to form the Hessian matrix.

5. Evaluate the eigenvalues of the Hessian matrix of the function $v(x, y) = x^4 + 2x^2y^2 + y^4$ at the point $(0, 0)$.

6. Apply Newton's Method on the function $w(x) = x^3 - 3x + 1$ starting at $x^{(0)} = 1$. Perform one iteration to find $x^{(1)}$.

Answers 3

1. Given a function $f(x, y) = 3x^2y + xy^2 + y^3$, compute the Hessian matrix:

   Solution:

   First compute the first-order partial derivatives:
   $$f'_x = 6xy + y^2, \quad f'_y = 3x^2 + 2xy + 3y^2$$

   Now compute the second-order partial derivatives:
   $$\frac{\partial^2 f}{\partial x^2} = 6y, \quad \frac{\partial^2 f}{\partial y^2} = 2x + 6y, \quad \frac{\partial^2 f}{\partial x \partial y} = \frac{\partial^2 f}{\partial y \partial x} = 6x + 2y$$

   Thus, the Hessian is:
   $$\nabla^2 f(x, y) = \begin{bmatrix} 6y & 6x + 2y \\ 6x + 2y & 2x + 6y \end{bmatrix}$$

2. Determine if the function $g(x, y) = x^2 + 4xy + 4y^2$ is convex by analyzing its Hessian matrix.

   Solution:
   $$g'_x = 2x + 4y, \quad g'_y = 4x + 8y$$
   $$\frac{\partial^2 g}{\partial x^2} = 2, \quad \frac{\partial^2 g}{\partial y^2} = 8, \quad \frac{\partial^2 g}{\partial x \partial y} = \frac{\partial^2 g}{\partial y \partial x} = 4$$
   $$\nabla^2 g(x, y) = \begin{bmatrix} 2 & 4 \\ 4 & 8 \end{bmatrix}$$

   Compute the determinant: $\det(\nabla^2 g) = (2)(8) - (4)(4) = 16 - 16 = 0$

   The determinant is non-negative and the diagonal entries are positive, indicate that g is convex.

3. For the function $h(x, y) = x^3 + 3xy + y^3$, find the critical points and classify them using the Hessian matrix.

   Solution:

   Critical points are where the gradient is zero:

   $$h'_x = 3x^2 + 3y, \quad h'_y = 3x + 3y^2$$

   Set the gradient to zero: $3x^2 + 3y = 0$ and $3x + 3y^2 = 0$

   $$\Rightarrow y = -x^2, \quad 3x - 3x^4 = 0$$

   $$\Rightarrow 3x(1 - x^3) = 0 \quad \Rightarrow x = 0, x^3 = 1$$

   At $x = 0, y = 0$. At $x = 1, y = -1$.

   Hessian matrix:

   $$\frac{\partial^2 h}{\partial x^2} = 6x, \quad \frac{\partial^2 h}{\partial y^2} = 6y, \quad \frac{\partial^2 h}{\partial x \partial y} = \frac{\partial^2 h}{\partial y \partial x} = 3$$

   $$\nabla^2 h(x, y) = \begin{bmatrix} 6x & 3 \\ 3 & 6y \end{bmatrix}$$

   At (0,0): $\begin{bmatrix} 0 & 3 \\ 3 & 0 \end{bmatrix}$.

   Determinant $= 0 - 9 = -9 \Rightarrow$ Saddle point

   At (1,-1): $\begin{bmatrix} 6 & 3 \\ 3 & -6 \end{bmatrix}$

   Determinant $= -36 - 9 = -45 \Rightarrow$ Saddle point

4. Consider the function $u(x, y, z) = x^2 y + yz^2 + xz$. Calculate the second-order partial derivatives to form the Hessian matrix.

   Solution:

   $$u'_x = 2xy + z, \quad u'_y = x^2 + z^2, \quad u'_z = 2yz + x$$

   Now compute the second-order partial derivatives:

   $$\frac{\partial^2 u}{\partial x^2} = 2y, \quad \frac{\partial^2 u}{\partial y^2} = 0, \quad \frac{\partial^2 u}{\partial z^2} = 2y$$

   $$\frac{\partial^2 u}{\partial x \partial y} = 2x, \quad \frac{\partial^2 u}{\partial x \partial z} = 1, \quad \frac{\partial^2 u}{\partial y \partial z} = 2z$$

   Thus, the Hessian is:

   $$\nabla^2 u(x, y, z) = \begin{bmatrix} 2y & 2x & 1 \\ 2x & 0 & 2z \\ 1 & 2z & 2y \end{bmatrix}$$

5. Evaluate the eigenvalues of the Hessian matrix of the function $v(x, y) = x^4 + 2x^2 y^2 + y^4$ at the point $(0, 0)$.

   Solution:

   $$v'_x = 4x^3 + 4xy^2, \quad v'_y = 4y^3 + 4x^2 y$$

   $$\frac{\partial^2 v}{\partial x^2} = 12x^2 + 4y^2, \quad \frac{\partial^2 v}{\partial y^2} = 12y^2 + 4x^2$$

   $$\frac{\partial^2 v}{\partial x \partial y} = \frac{\partial^2 v}{\partial y \partial x} = 8xy$$

   Evaluate at $(0, 0) : \nabla^2 v(0, 0) = \begin{bmatrix} 0 & 0 \\ 0 & 0 \end{bmatrix}$

   Eigenvalues of the zero matrix are $\lambda_1 = 0, \lambda_2 = 0$.

6. Apply Newton's Method on the function $w(x) = x^3 - 3x + 1$ starting at $x^{(0)} = 1$. Perform one iteration to find $x^{(1)}$.

   Solution:

$$\text{Compute the derivatives: } w'(x) = 3x^2 - 3, \quad w''(x) = 6x$$

$$\text{At } x^{(0)} = 1 :$$

$$w'(1) = 3(1)^2 - 3 = 0, \quad w''(1) = 6(1) = 6$$

$$\text{Newton's update step:}$$

$$x^{(1)} = 1 - \frac{w'(1)}{w''(1)} = 1 - \frac{0}{6} = 1$$

$$\text{Thus, } x^{(1)} = 1.$$

Chapter 10

Taylor Series and Function Approximation

Practice Problems 1

1. Derive the Taylor series for the function $f(x) = \cos(x)$ about $a = 0$ up to the fourth-order term.

2. Compute the Taylor series approximation of $f(x) = \ln(1+x)$ around $a = 0$ up to the third-order term.

3. For the function $f(x) = e^{-x}$, determine the second-order Taylor series expansion centered at $a = 1$.

4. Use the Taylor series to approximate the value of $\sin(0.1)$ using expansion about $a = 0$ including terms up to x^5.

5. Analyze the error in using the second-order Taylor polynomial of $f(x) = \sqrt{1+x}$ about $a = 0$ to approximate $f(0.1)$.

6. Consider the function $f(x) = x^2 e^x$. Find the first three non-zero terms of its Taylor series expanded about $a = 0$.

Answers 1

1. Derive the Taylor series for the function $f(x) = \cos(x)$ about $a = 0$ up to the fourth-order term.

   Solution:
   $$f(x) = \cos(x), \quad f^{(0)}(x) = \cos(x), \quad f^{(n)}(a) \text{ for } a = 0 :$$
   $$f(0) = 1, \quad f'(x) = -\sin(x) \quad \Rightarrow \quad f'(0) = 0,$$
   $$f''(x) = -\cos(x) \quad \Rightarrow \quad f''(0) = -1,$$
   $$f'''(x) = \sin(x) \quad \Rightarrow \quad f'''(0) = 0,$$
   $$f^{(4)}(x) = \cos(x) \quad \Rightarrow \quad f^{(4)}(0) = 1.$$

   Therefore, the Taylor series is:
   $$f(x) = 1 - \frac{x^2}{2!} + \frac{x^4}{4!} + \cdots.$$

96

2. Compute the Taylor series approximation of $f(x) = \ln(1+x)$ around $a = 0$ up to the third-order term.

   Solution:
   $$f(x) = \ln(1+x), \quad \text{find derivatives at } x = 0:$$
   $$f'(x) = \frac{1}{1+x}, \quad f'(0) = 1,$$
   $$f''(x) = -\frac{1}{(1+x)^2}, \quad f''(0) = -1,$$
   $$f'''(x) = \frac{2}{(1+x)^3}, \quad f'''(0) = 2.$$

   The Taylor series up to third-order is:
   $$f(x) = x - \frac{x^2}{2} + \frac{x^3}{3} + \cdots.$$

3. For the function $f(x) = e^{-x}$, determine the second-order Taylor series expansion centered at $a = 1$.

   Solution:
   $$f(x) = e^{-x}, \quad f^{(n)}(x) = (-1)^n e^{-x}, \quad \text{evaluate at } x = 1:$$
   $$f(1) = e^{-1}, \quad f'(1) = -e^{-1}, \quad f''(1) = e^{-1}.$$

   The second-order expansion about $a = 1$ is:
   $$f(x) = e^{-1} - (x-1)e^{-1} + \frac{(x-1)^2}{2}e^{-1}.$$

   Therefore, the expansion is:
   $$f(x) = e^{-1}\left(1 - (x-1) + \frac{(x-1)^2}{2}\right).$$

4. Use the Taylor series to approximate the value of $\sin(0.1)$ using expansion about $a = 0$ including terms up to x^5.

   Solution:
   $$\sin(x) \approx x - \frac{x^3}{3!} + \frac{x^5}{5!} + \cdots,$$
   $$\sin(0.1) \approx 0.1 - \frac{0.1^3}{6} + \frac{0.1^5}{120}.$$
   $$= 0.1 - 0.0001667 + 0.000000833,$$
   $$\approx 0.0998333.$$

5. Analyze the error in using the second-order Taylor polynomial of $f(x) = \sqrt{1+x}$ about $a = 0$ to approximate $f(0.1)$.

   Solution:
   $$f(x) = (1+x)^{1/2}, \quad f'(x) = \frac{1}{2}(1+x)^{-1/2}, \quad f''(x) = -\frac{1}{4}(1+x)^{-3/2}.$$

   The second-order Taylor polynomial about $a = 0$:
   $$f(x) \approx 1 + \frac{x}{2} - \frac{x^2}{8}.$$

At $x = 0.1$:
$$f(0.1) \approx 1 + \frac{0.1}{2} - \frac{0.01}{8} = 1.049375.$$

The actual value is:
$$\sqrt{1.1} \approx 1.04881.$$

Error = Approximation - Actual = $1.049375 - 1.04881 = 0.000565$.

6. Consider the function $f(x) = x^2 e^x$. Find the first three non-zero terms of its Taylor series expanded about $a = 0$.

   Solution:
   $$f(x) = x^2 e^x, \quad f'(x) = (x^2 + 2x)e^x, \quad f''(x) = (x^2 + 4x + 2)e^x.$$

   Evaluation for first few terms:
   $$f(0) = 0, \quad f'(0) = 0, \quad f''(0) = 2.$$

   The Taylor series expansion is:
   $$f(x) = 0 + 0 + \frac{2}{2!}x^2 + \cdots = x^2 + \cdots.$$

   Seeking more terms:
   $$f'''(x) = ((x^2 + 6x + 6)e^x) \Rightarrow f'''(0) = 6, \quad f(x) = x^2 + x^3 + \cdots.$$

Practice Problems 2

1. Derive the Taylor series expansion for the function $f(x) = \log(1 + x)$ about $a = 0$ up to the fourth order.

2. For the function $f(x) = \cos(x)$, find the Taylor series expansion centered at $a = 0$ up to the sixth order and evaluate its approximation at $x = \frac{\pi}{4}$.

3. Apply the Taylor series approximation to compute the value of e^x at $x = 0.1$ using the expansion up to the third order.

4. Determine the error bound for the Taylor series approximation of $\sin(x)$ up to the second order at $x = 0.2$.

5. Use a Taylor series expansion to approximate the value of $\sqrt{1.1}$ using the function $f(x) = \sqrt{1+x}$ centered at $a = 0$ up to the second order.

6. For the function $f(x) = \tanh(x)$, develop the third-order Taylor polynomial at $a = 0$ and calculate its value at $x = 0.2$.

Answers 2

1. Derive the Taylor series expansion for the function $f(x) = \log(1 + x)$ about $a = 0$ up to the fourth order.

   Solution:

   The Taylor series for the function is given by:

   $$f(x) = \log(1 + x) = x - \frac{x^2}{2} + \frac{x^3}{3} - \frac{x^4}{4} + \cdots$$

   Therefore, up to the fourth order:

   $$f(x) \approx x - \frac{x^2}{2} + \frac{x^3}{3} - \frac{x^4}{4}$$

2. For the function $f(x) = \cos(x)$, find the Taylor series expansion centered at $a = 0$ up to the sixth order and evaluate its approximation at $x = \frac{\pi}{4}$.

   Solution:

99

The Taylor series for $\cos(x)$ is:

$$\cos(x) = 1 - \frac{x^2}{2!} + \frac{x^4}{4!} - \frac{x^6}{6!} + \cdots$$

Up to the sixth order:

$$\cos(x) \approx 1 - \frac{x^2}{2} + \frac{x^4}{24} - \frac{x^6}{720}$$

Evaluating at $x = \frac{\pi}{4}$:

$$\cos\left(\frac{\pi}{4}\right) \approx 1 - \frac{\left(\frac{\pi}{4}\right)^2}{2} + \frac{\left(\frac{\pi}{4}\right)^4}{24} - \frac{\left(\frac{\pi}{4}\right)^6}{720}$$

$$\approx 1 - \frac{\pi^2}{32} + \frac{\pi^4}{1536} - \frac{\pi^6}{184320}$$

3. Apply the Taylor series approximation to compute the value of e^x at $x = 0.1$ using the expansion up to the third order.
   Solution:

   The Taylor series for e^x is:

   $$e^x = 1 + x + \frac{x^2}{2!} + \frac{x^3}{3!} + \cdots$$

   Up to the third order:

   $$e^x \approx 1 + x + \frac{x^2}{2} + \frac{x^3}{6}$$

   Evaluating at $x = 0.1$:

   $$e^{0.1} \approx 1 + 0.1 + \frac{0.1^2}{2} + \frac{0.1^3}{6}$$

   $$= 1 + 0.1 + 0.005 + 0.0001667 = 1.105167$$

4. Determine the error bound for the Taylor series approximation of $\sin(x)$ up to the second order at $x = 0.2$.
   Solution:

   The error bound for the Taylor series is given by:

   $$R_n(x) = \frac{f^{(n+1)}(\xi)}{(n+1)!}(x - a)^{n+1}$$

   For the function $f(x) = \sin(x)$ up to second order:

   $$R_2(x) = \frac{f^{(3)}(\xi)}{3!}x^3 = \frac{\cos(\xi)}{6}x^3$$

   Using $\cos(\xi) \leq 1$ for all ξ, we have:

   $$R_2(0.2) \leq \frac{1}{6} \times 0.2^3 = \frac{0.008}{6} = 0.001333$$

5. Use a Taylor series expansion to approximate the value of $\sqrt{1.1}$ using the function $f(x) = \sqrt{1 + x}$ centered at $a = 0$ up to the second order.
   Solution:

   The function $f(x) = \sqrt{1 + x}$ can be represented as:

   $$\sqrt{1 + x} \approx 1 + \frac{x}{2} - \frac{x^2}{8}$$

100

Evaluating at $x = 0.1$:

$$\sqrt{1.1} \approx 1 + \frac{0.1}{2} - \frac{0.1^2}{8}$$

$$= 1 + 0.05 - 0.00125 = 1.04875$$

6. For the function $f(x) = \tanh(x)$, develop the third-order Taylor polynomial at $a = 0$ and calculate its value at $x = 0.2$.

   Solution:

   The Taylor series for $\tanh(x)$ at $a = 0$ is:

   $$\tanh(x) \approx x - \frac{x^3}{3}$$

   Evaluating at $x = 0.2$:

   $$\tanh(0.2) \approx 0.2 - \frac{0.2^3}{3}$$

   $$= 0.2 - \frac{0.008}{3} = 0.2 - 0.002667 = 0.197333$$

Practice Problems 3

1. Derive the Taylor series expansion for the function $f(x) = \cos(x)$ around $a = 0$ up to the fourth-order term.

2. Compute the Taylor series approximation of the function $f(x) = \ln(1 + x)$ about $a = 0$ up to the third-order term.

3. Consider the function $f(x) = e^{x^2}$. Find the first three non-zero terms of the Taylor series expansion around $x = 0$.

101

4. Use Taylor series to approximate $\sin(0.1)$ using the series around $a = 0$ up to the third-order term.

5. Determine the remainder term $R_2(x)$ for the second-order Taylor polynomial of $f(x) = \sqrt{1+x}$ around $a = 0$.

6. Apply the Taylor series expansion to approximate $e^{-0.5}$ using terms up to $n = 3$.

Answers 3

1. Derive the Taylor series expansion for the function $f(x) = \cos(x)$ around $a = 0$ up to the fourth-order term.
   Solution:
   $$f(x) = \cos(x), \quad f(0) = 1$$

   The derivatives are:
   $$f'(x) = -\sin(x), \quad f'(0) = 0$$
   $$f''(x) = -\cos(x), \quad f''(0) = -1$$
   $$f'''(x) = \sin(x), \quad f'''(0) = 0$$
   $$f^{(4)}(x) = \cos(x), \quad f^{(4)}(0) = 1$$

   Thus, the Taylor series up to fourth-order term is:

   $$\cos(x) \approx 1 - \frac{x^2}{2!} + \frac{x^4}{4!}$$

2. Compute the Taylor series approximation of the function $f(x) = \ln(1 + x)$ about $a = 0$ up to the third-order term.

   Solution:
   $$f(x) = \ln(1 + x), \quad f(0) = 0$$

   The derivatives are:
   $$f'(x) = \frac{1}{1+x}, \quad f'(0) = 1$$
   $$f''(x) = -\frac{1}{(1+x)^2}, \quad f''(0) = -1$$
   $$f'''(x) = \frac{2}{(1+x)^3}, \quad f'''(0) = 2$$

   Thus, the Taylor series up to third-order term is:
   $$\ln(1 + x) \approx x - \frac{x^2}{2} + \frac{x^3}{3}$$

3. Consider the function $f(x) = e^{x^2}$. Find the first three non-zero terms of the Taylor series expansion around $x = 0$.

   Solution:
   $$f(x) = e^{x^2}, \quad f(0) = 1$$

   The derivatives are:
   $$f'(x) = 2xe^{x^2}, \quad f'(0) = 0$$
   $$f''(x) = (4x^2 + 2)e^{x^2}, \quad f''(0) = 2$$
   $$f'''(x) = (8x^3 + 12x)e^{x^2}, \quad f'''(0) = 0$$
   $$f^{(4)}(x) = (16x^4 + 48x^2 + 12)e^{x^2}, \quad f^{(4)}(0) = 12$$

   Thus, the first three non-zero terms are:
   $$e^{x^2} \approx 1 + \frac{x^2}{1!} + \frac{12x^4}{4!}$$

4. Use Taylor series to approximate $\sin(0.1)$ using the series around $a = 0$ up to the third-order term.

   Solution:
   $$\sin(x) \approx x - \frac{x^3}{3!}$$

   Substitute $x = 0.1$:
   $$\sin(0.1) \approx 0.1 - \frac{0.1^3}{6}$$
   $$= 0.1 - \frac{0.001}{6} = 0.1 - 0.0001667 \approx 0.0998333$$

5. Determine the remainder term $R_2(x)$ for the second-order Taylor polynomial of $f(x) = \sqrt{1 + x}$ around $a = 0$.

   Solution: For the function $f(x) = \sqrt{1 + x}$:
   $$R_2(x) = \frac{f^{(3)}(\xi)}{3!}(x)^3$$

   Calculating derivatives:
   $$f'(x) = \frac{1}{2\sqrt{1+x}}, \quad f''(x) = -\frac{1}{4(1+x)^{3/2}}, \quad f'''(x) = \frac{3}{8(1+x)^{5/2}}$$

   At $x = 0$, $f'''(0) = \frac{3}{8}$. So, the remainder is:
   $$R_2(x) = \frac{3}{48}(x)^3 = \frac{1}{16}x^3$$

103

6. Apply the Taylor series expansion to approximate $e^{-0.5}$ using terms up to $n = 3$.

Solution:

$$e^x \approx 1 + x + \frac{x^2}{2!} + \frac{x^3}{3!}$$

Substitute $x = -0.5$:

$$e^{-0.5} \approx 1 - 0.5 + \frac{(-0.5)^2}{2} + \frac{(-0.5)^3}{6}$$

$$= 1 - 0.5 + \frac{0.25}{2} - \frac{0.125}{6}$$

$$= 1 - 0.5 + 0.125 - 0.0208333$$

$$\approx 0.6041667$$

Chapter 11

Multivariable Taylor Series

Practice Problems 1

1. Verify the multivariable Taylor series expansion of a quadratic function $f(x,y) = 3x^2 + 2xy + y^2 + 4x + y + 5$ around the point $(0,0)$ using the general formula provided in the chapter.

2. Consider the function $g(x,y) = e^x \sin(y)$. Find the first-order Taylor series approximation around $(0,0)$ and evaluate it at the point $(0.1, 0.1)$.

3. Given the function $h(x,y,z) = x^2 y + yz^3$, compute the gradient vector ∇h at the point $(1,1,1)$.

4. For the function $f(x,y) = \ln(x^2 + y^2)$, calculate the Hessian matrix at the point $(1,0)$.

5. Using the function $k(x, y) = x^3 + 3xy + y^3$, determine the second-order approximation around the point $(1, 1)$.

6. Apply Newton's method for optimization to find the critical point of the function $f(x, y) = x^2 + y^2 - 4x - 6y + 13$. Use the initial guess $(2, 3)$.

Answers 1

1. **Solution:** Verify the series expansion of $f(x, y) = 3x^2 + 2xy + y^2 + 4x + y + 5$ around $(0, 0)$:

$$f(0, 0) = 5.$$

First derivatives:

$$\frac{\partial f}{\partial x} = 6x + 2y + 4, \quad \frac{\partial f}{\partial y} = 2x + 2y + 1.$$

Evaluate at $(0, 0)$:

$$\left.\frac{\partial f}{\partial x}\right|_{(0,0)} = 4, \quad \left.\frac{\partial f}{\partial y}\right|_{(0,0)} = 1.$$

Second derivatives (Hessian components):

$$\frac{\partial^2 f}{\partial x^2} = 6, \quad \frac{\partial^2 f}{\partial x \partial y} = 2, \quad \frac{\partial^2 f}{\partial y^2} = 2.$$

Hessian at $(0, 0)$ is:

$$H = \begin{bmatrix} 6 & 2 \\ 2 & 2 \end{bmatrix}.$$

Using the formula:

$$f(x, y) \approx 5 + 4x + 1y + \frac{1}{2} \begin{bmatrix} x & y \end{bmatrix} \begin{bmatrix} 6 & 2 \\ 2 & 2 \end{bmatrix} \begin{bmatrix} x \\ y \end{bmatrix}.$$

Simplifying:

$$\approx 5 + 4x + y + 3x^2 + 2xy + y^2.$$

2. **Solution:** Find the first-order approximation of $g(x, y) = e^x \sin(y)$ at $(0, 0)$:

$$g(0, 0) = e^0 \sin(0) = 0.$$

Derivatives:

$$\frac{\partial g}{\partial x} = e^x \sin(y), \quad \frac{\partial g}{\partial y} = e^x \cos(y).$$

Evaluating at $(0, 0)$:

$$\left.\frac{\partial g}{\partial x}\right|_{(0,0)} = 0, \quad \left.\frac{\partial g}{\partial y}\right|_{(0,0)} = 1.$$

First-order approximation:

$$g(x, y) \approx 0 + 0 \cdot x + 1 \cdot y = y.$$

Evaluating at $(0.1, 0.1)$:

$$g(0.1, 0.1) \approx 0.1.$$

3. **Solution:** Compute the gradient of $h(x, y, z) = x^2 y + yz^3$ at $(1, 1, 1)$:

$$\nabla h = \left(\frac{\partial h}{\partial x}, \frac{\partial h}{\partial y}, \frac{\partial h}{\partial z}\right).$$

Derivatives:

$$\frac{\partial h}{\partial x} = 2xy, \quad \frac{\partial h}{\partial y} = x^2 + z^3, \quad \frac{\partial h}{\partial z} = 3yz^2.$$

Evaluate at $(1, 1, 1)$:

$$\nabla h(1, 1, 1) = (2 \cdot 1 \cdot 1, 1^2 + 1^3, 3 \cdot 1 \cdot 1^2) = (2, 2, 3).$$

4. **Solution:** Calculate the Hessian matrix for $f(x, y) = \ln(x^2 + y^2)$ at $(1, 0)$:

$$f_x = \frac{2x}{x^2 + y^2}, \quad f_y = \frac{2y}{x^2 + y^2}.$$

Derivatives of these:

$$f_{xx} = \frac{2(y^2 - x^2)}{(x^2 + y^2)^2}, \quad f_{yy} = \frac{2(x^2 - y^2)}{(x^2 + y^2)^2}, \quad f_{xy} = f_{yx} = \frac{-4xy}{(x^2 + y^2)^2}.$$

Evaluate at $(1, 0)$:

$$f_{xx}(1, 0) = 2, \quad f_{yy}(1, 0) = -2, \quad f_{xy}(1, 0) = 0.$$

Hessian:

$$H = \begin{bmatrix} 2 & 0 \\ 0 & -2 \end{bmatrix}.$$

5. **Solution:** Second-order approximation of $k(x, y) = x^3 + 3xy + y^3$ around $(1, 1)$:

$$k(1, 1) = 1 + 3 + 1 = 5.$$

Gradients:

$$\frac{\partial k}{\partial x} = 3x^2 + 3y, \quad \frac{\partial k}{\partial y} = 3x + 3y^2.$$

Evaluate at $(1, 1)$:

$$\nabla k(1, 1) = (6, 6).$$

Hessian:

$$H = \begin{bmatrix} 6x & 3 \\ 3 & 6y \end{bmatrix}.$$

Evaluate at $(1, 1)$:

$$H = \begin{bmatrix} 6 & 3 \\ 3 & 6 \end{bmatrix}.$$

Second-order approximation:

$$k(x, y) \approx 5 + 6(x - 1) + 6(y - 1) + \frac{1}{2} \begin{bmatrix} x - 1 & y - 1 \end{bmatrix} \begin{bmatrix} 6 & 3 \\ 3 & 6 \end{bmatrix} \begin{bmatrix} x - 1 \\ y - 1 \end{bmatrix}.$$

6. **Solution:** Find the critical point of $f(x, y) = x^2 + y^2 - 4x - 6y + 13$ using Newton's method:

$$\nabla f = (2x - 4, 2y - 6).$$

Hessian:

$$H = \begin{bmatrix} 2 & 0 \\ 0 & 2 \end{bmatrix}.$$

Starting from $(2, 3)$:

$$\mathbf{x}_{k+1} = \mathbf{x}_k - H^{-1} \nabla f(\mathbf{x}_k).$$

Calculate:

$$\nabla f(2, 3) = (0, 0).$$

Since the gradient is zero, the initial guess is already the critical point. Hence:

$$(2, 3) \text{ is the critical point.}$$

Practice Problems 2

1. Find the first-order Taylor series approximation of the function, $f(x, y) = e^x \sin(y)$, around the point $(0, 0)$.

2. Compute the Hessian matrix for the function $f(x, y) = x^3 + xy + y^3$ at the point $(1, -1)$.

3. Use the second-order Taylor series to approximate the function $f(x, y) = \ln(1 + x + y)$ near the point $(0, 0)$.

4. Verify if the quadratic function $f(x, y) = x^2 - 4xy + 5y^2$ has local minima, maxima, or saddle points.

5. Consider the function $f(x, y) = x^2 y + 3xy - y^2$. Perform a first-order approximation around the point $(1, 2)$.

6. Apply Newton's method to find one iteration update for the function $f(x, y) = x^4 + y^4 - 4xy$ starting from the point $(1, 1)$.

Answers 2

1. For the first-order Taylor series approximation of $f(x, y) = e^x \sin(y)$ around $(0, 0)$,

First, compute the partial derivatives:

$$\frac{\partial f}{\partial x} = e^x \sin(y), \quad \frac{\partial f}{\partial y} = e^x \cos(y)$$

Evaluate these at $(0, 0)$: $\quad \frac{\partial f}{\partial x}(0, 0) = \sin(0) = 0, \quad \frac{\partial f}{\partial y}(0, 0) = 1$

The first-order approximation is: $\quad f(x, y) \approx f(0, 0) + 0 \cdot x + 1 \cdot y = y$

2. For the Hessian of $f(x, y) = x^3 + xy + y^3$ at $(1, -1)$,

Compute the second partial derivatives:

$$\frac{\partial^2 f}{\partial x^2} = 6x, \quad \frac{\partial^2 f}{\partial y^2} = 6y, \quad \frac{\partial^2 f}{\partial x \partial y} = 1$$

Evaluate these at $(1, -1)$: $\quad \dfrac{\partial^2 f}{\partial x^2}(1, -1) = 6, \quad \dfrac{\partial^2 f}{\partial y^2}(1, -1) = -6, \quad \dfrac{\partial^2 f}{\partial x \partial y}(1, -1) = 1$

The Hessian matrix is: $\quad H_f(1, -1) = \begin{bmatrix} 6 & 1 \\ 1 & -6 \end{bmatrix}$

3. For the second-order Taylor series approximation of $f(x, y) = \ln(1 + x + y)$ around $(0, 0)$,

Compute the partial derivatives:

$$\frac{\partial f}{\partial x} = \frac{1}{1 + x + y}, \quad \frac{\partial f}{\partial y} = \frac{1}{1 + x + y}$$

$$\frac{\partial^2 f}{\partial x^2} = -\frac{1}{(1 + x + y)^2}, \quad \frac{\partial^2 f}{\partial y^2} = -\frac{1}{(1 + x + y)^2}, \quad \frac{\partial^2 f}{\partial x \partial y} = -\frac{1}{(1 + x + y)^2}$$

Evaluating at $(0, 0)$: $\quad \dfrac{\partial f}{\partial x}(0, 0) = 1, \quad$ and similarly, $\dfrac{\partial^2 f}{\partial x^2}(0, 0) = -1$

The approximation is: $\quad f(x, y) \approx \ln(1) + 1 \cdot x + 1 \cdot y - \dfrac{1}{2}(x^2 + 2xy + y^2) = x + y - \dfrac{1}{2}(x^2 + 2xy + y^2)$

4. For $f(x, y) = x^2 - 4xy + 5y^2$,

Compute the Hessian matrix:

$$\frac{\partial^2 f}{\partial x^2} = 2, \quad \frac{\partial^2 f}{\partial y^2} = 10, \quad \frac{\partial^2 f}{\partial x \partial y} = -4$$

$$H_f = \begin{bmatrix} 2 & -4 \\ -4 & 10 \end{bmatrix}$$

Evaluating the eigenvalues: $\quad \det(H - \lambda I) = \begin{vmatrix} 2 - \lambda & -4 \\ -4 & 10 - \lambda \end{vmatrix} = \lambda^2 - 12\lambda + 12$

The eigenvalues are $\lambda_1 = 2, \lambda_2 = 10$. Since both are positive, it is a local minimum.

5. For the function $f(x, y) = x^2 y + 3xy - y^2$ with a first-order approximation around $(1, 2)$,

$$\frac{\partial f}{\partial x} = 2xy + 3y, \quad \frac{\partial f}{\partial y} = x^2 + 3x - 2y$$

Evaluate at $(1, 2)$: $\quad \dfrac{\partial f}{\partial x}(1, 2) = 10, \quad \dfrac{\partial f}{\partial y}(1, 2) = 0$

The approximation is: $\quad f(x, y) \approx f(1, 2) + 10(x - 1) = 3 + 10(x - 1)$

6. Apply Newton's method to $f(x, y) = x^4 + y^4 - 4xy$,

$$\nabla f = \begin{bmatrix} 4x^3 - 4y \\ 4y^3 - 4x \end{bmatrix}, \quad H_f = \begin{bmatrix} 12x^2 & -4 \\ -4 & 12y^2 \end{bmatrix}$$

At $(1, 1)$: $\quad \nabla f(1, 1) = \begin{bmatrix} 0 \\ 0 \end{bmatrix}, \quad H_f(1, 1) = \begin{bmatrix} 12 & -4 \\ -4 & 12 \end{bmatrix}$

Solve $H_f(1, 1)\Delta \mathbf{x} = \nabla f(1, 1) \Rightarrow \Delta \mathbf{x} = \begin{bmatrix} 0 \\ 0 \end{bmatrix}$

No update occurs because the gradient is zero at the starting point.

Practice Problems 3

1. Compute the first-order Taylor series approximation of the function $f(x, y) = e^{x+y}$ at the point $(0, 0)$.

2. Determine the second-order Taylor series expansion for the function $f(x, y) = x^2 - xy + y^2$ around the point $(1, 1)$.

3. Find the Hessian matrix of $f(x, y) = \sin(x)y + \cos(y)x$ and evaluate it at the point $(\pi, 0)$.

4. Using Taylor series, approximate $f(x, y) = \ln(x^2 + y^2)$ near the point $(1, 1)$ up to the second order.

5. Describe the role of the Hessian matrix in determining the local extremum of a function f at a point.

6. Use Newton's method to find an iteration formula for the function $f(x, y) = x^2 + y^2 + 2xy - 4x$ starting at a point close to the origin.

Answers 3

1. Compute the first-order Taylor series approximation of the function $f(x, y) = e^{x+y}$ at the point $(0, 0)$.
   Solution:
   The first-order Taylor series is given by:

   $$f(x, y) \approx f(0, 0) + \frac{\partial f}{\partial x}(0, 0)x + \frac{\partial f}{\partial y}(0, 0)y$$

   Calculate $f(0, 0) = e^{0+0} = 1$.

   $$\frac{\partial f}{\partial x} = e^{x+y}, \quad \frac{\partial f}{\partial y} = e^{x+y}$$

   At $(0, 0)$:

   $$\frac{\partial f}{\partial x}(0, 0) = 1, \quad \frac{\partial f}{\partial y}(0, 0) = 1$$

   Taylor series approximation:

   $$f(x, y) \approx 1 + x + y$$

2. Determine the second-order Taylor series expansion for $f(x, y) = x^2 - xy + y^2$ around $(1, 1)$.
   Solution:
   Compute: $f(1, 1) = 1^2 - 1 \cdot 1 + 1^2 = 1$.
   Partial derivatives:

   $$\frac{\partial f}{\partial x} = 2x - y, \quad \frac{\partial f}{\partial y} = -x + 2y$$

   At $(1, 1)$:

   $$\frac{\partial f}{\partial x}(1, 1) = 1, \quad \frac{\partial f}{\partial y}(1, 1) = 1$$

   Second derivatives for Hessian:

   $$\frac{\partial^2 f}{\partial x^2} = 2, \quad \frac{\partial^2 f}{\partial y^2} = 2, \quad \frac{\partial^2 f}{\partial x \partial y} = -1$$

   Hessian H at $(1, 1)$ is:

   $$H = \begin{bmatrix} 2 & -1 \\ -1 & 2 \end{bmatrix}$$

   Second-order Taylor expansion:

   $$f(x, y) \approx 1 + (1)(x - 1) + (1)(y - 1) + \frac{1}{2} \begin{bmatrix} x - 1 & y - 1 \end{bmatrix} H \begin{bmatrix} x - 1 \\ y - 1 \end{bmatrix}$$

112

Simplify:

$$= 1 + x - 1 + y - 1 + \frac{1}{2}\left[2(x-1)^2 - 2(x-1)(y-1) + 2(y-1)^2\right]$$

$$= x + y - 1 + (x-1)^2 - (x-1)(y-1) + (y-1)^2$$

3. Find the Hessian matrix of $f(x,y) = \sin(x)y + \cos(y)x$ and evaluate at $(\pi, 0)$.
   Solution:
   Compute partial derivatives:

   $$\frac{\partial f}{\partial x} = \cos(x)y + \cos(y), \quad \frac{\partial f}{\partial y} = \sin(x) - x\sin(y)$$

   Second-order partial derivatives:

   $$\frac{\partial^2 f}{\partial x^2} = -\sin(x)y, \quad \frac{\partial^2 f}{\partial y^2} = -x\cos(y)$$

   $$\frac{\partial^2 f}{\partial x \partial y} = \cos(x), \quad \frac{\partial^2 f}{\partial y \partial x} = \cos(x)$$

   Hessian H is:

   $$H = \begin{bmatrix} -\sin(x)y & \cos(x) \\ \cos(x) & -x\cos(y) \end{bmatrix}$$

   At $(\pi, 0)$:

   $$H(\pi, 0) = \begin{bmatrix} 0 & -1 \\ -1 & -\pi \end{bmatrix}$$

4. Using Taylor series, approximate $f(x,y) = \ln(x^2 + y^2)$ near $(1,1)$ up to the second order.
   Solution:
   Compute $f(1,1) = \ln(1^2 + 1^2) = \ln(2)$.
   Partial derivatives:
   $$\frac{\partial f}{\partial x} = \frac{2x}{x^2 + y^2}, \quad \frac{\partial f}{\partial y} = \frac{2y}{x^2 + y^2}$$

   At $(1,1)$:
   $$\frac{\partial f}{\partial x}(1,1) = 1, \quad \frac{\partial f}{\partial y}(1,1) = 1$$

   Second partial derivatives:
   $$\frac{\partial^2 f}{\partial x^2} = \frac{2(y^2 - x^2)}{(x^2 + y^2)^2}, \quad \frac{\partial^2 f}{\partial y^2} = \frac{2(x^2 - y^2)}{(x^2 + y^2)^2}, \quad \frac{\partial^2 f}{\partial x \partial y} = \frac{-4xy}{(x^2 + y^2)^2}$$

   At $(1,1)$:
   $$\frac{\partial^2 f}{\partial x^2}(1,1) = 0, \quad \frac{\partial^2 f}{\partial y^2}(1,1) = 0, \quad \frac{\partial^2 f}{\partial x \partial y}(1,1) = -1$$

   Taylor expansion:
   $$f(x,y) \approx \ln(2) + x - 1 + y - 1 - \frac{1}{2}(x-1)(y-1)$$

5. Describe the role of the Hessian matrix in determining the local extremum of a function f at a point.
   Answer:
   The Hessian matrix, composed of second-order partial derivatives, helps ascertain the nature of a critical point, $\mathbf{a}$, of a differentiable function f:

   - If the Hessian is positive definite at $\mathbf{a}$, f has a local minimum there.
   - If the Hessian is negative definite at $\mathbf{a}$, f has a local maximum there.
   - If the Hessian is indefinite, $\mathbf{a}$ is a saddle point.

113

- If it's singular, the test is inconclusive.

6. Use Newton's method to find an iteration formula for $f(x, y) = x^2 + y^2 + 2xy - 4x$.

   Solution:

   Newton's method updates using:

   $$\mathbf{x}_{k+1} = \mathbf{x}_k - H(f)^{-1} \nabla f(\mathbf{x}_k)$$

   Compute:

   $$\nabla f = \begin{bmatrix} 2x + 2y - 4 \\ 2x + 2y \end{bmatrix}$$

   Compute Hessian:

   $$H = \begin{bmatrix} 2 & 2 \\ 2 & 2 \end{bmatrix}$$

   At any point, $(\hat{x}, \hat{y})$, H^{-1}:

   $$H^{-1} = \begin{bmatrix} 0.5 & -0.5 \\ -0.5 & 0.5 \end{bmatrix}$$

   Iteration:

   $$\begin{bmatrix} x_{k+1} \\ y_{k+1} \end{bmatrix} = \begin{bmatrix} x_k \\ y_k \end{bmatrix} - \begin{bmatrix} 0.5 & -0.5 \\ -0.5 & 0.5 \end{bmatrix} \begin{bmatrix} 2x_k + 2y_k - 4 \\ 2x_k + 2y_k \end{bmatrix}$$

   Simplified:

   $$\begin{bmatrix} x_{k+1} \\ y_{k+1} \end{bmatrix} = \begin{bmatrix} x_k + 2 - x_k - y_k \\ y_k + 2 - x_k - y_k \end{bmatrix} = \begin{bmatrix} 2 - y_k \\ 2 - x_k \end{bmatrix}$$

   Repeat until convergence.

Chapter 12

Chain Rule in Multivariable Calculus

Practice Problems 1

1. Let $f : \mathbb{R}^2 \to \mathbb{R}^3$ be defined by $f(x, y) = (x^2 + y, xy, y^2)$ and $g : \mathbb{R}^3 \to \mathbb{R}$ by $g(u, v, w) = u + v^2 + w^3$. Find the derivative of the composite function $h(x, y) = g(f(x, y))$ at the point $(1, 2)$.

2. Evaluate the gradient of the function $h(x, y, z) = e^{x+y+z}$ along the direction of the vector $\mathbf{v} = (1, -1, 0)$ at the point $(0, -1, 2)$.

3. Consider the composite function $p(t, s) = \sin(u(t, s))$, where $u(t, s) = e^{ts}$. Use the chain rule to find the partial derivatives $\frac{\partial p}{\partial t}$ and $\frac{\partial p}{\partial s}$.

4. Find the Jacobian of the function $f(x, y) = (\ln(xy), x^2 - 4y)$ at the point $(2, 1)$.

5. Given $f(x, y) = x^y$ and $g(z) = \ln(z)$, compute the derivative of $h(x, y) = g(f(x, y))$ using the chain rule. Express your answer in terms of x and y.

6. In a neural network, let the activation of a neuron be described by $a = \sigma(wx + b)$, where $\sigma(z) = \frac{1}{1+e^{-z}}$. Calculate the derivative $\frac{da}{dw}$ using the chain rule.

Answers 1

1. Consider $f(x, y) = (x^2 + y, xy, y^2)$ and $g(u, v, w) = u + v^2 + w^3$. We aim to find $\nabla h(x, y)$ at $(1, 2)$.

   Solution:

   $$J_f(1, 2) = \begin{bmatrix} 2x & 1 \\ y & x \\ 0 & 2y \end{bmatrix} \Bigg|_{(x,y)=(1,2)} = \begin{bmatrix} 2 & 1 \\ 2 & 1 \\ 0 & 4 \end{bmatrix}$$

   $$J_g = \begin{bmatrix} 1 & 2v & 3w^2 \end{bmatrix} \Bigg|_{u=3, v=2, w=4} = \begin{bmatrix} 1 & 4 & 48 \end{bmatrix}$$

   $$J_h = J_g \cdot J_f(1, 2) = \begin{bmatrix} 1 & 4 & 48 \end{bmatrix} \cdot \begin{bmatrix} 2 & 1 \\ 2 & 1 \\ 0 & 4 \end{bmatrix} = \begin{bmatrix} 10 & 205 \end{bmatrix}$$

   Therefore, the derivative at $(1, 2)$ is $\begin{bmatrix} 10 & 205 \end{bmatrix}$.

116

2. Given $h(x, y, z) = e^{x+y+z}$, evaluate the gradient ∇h first:

   Solution:
   $$\nabla h = \left(\frac{\partial h}{\partial x}, \frac{\partial h}{\partial y}, \frac{\partial h}{\partial z} \right) = \left(e^{x+y+z}, e^{x+y+z}, e^{x+y+z} \right)$$

   At $(0, -1, 2)$, this becomes $(e^1, e^1, e^1) = (e, e, e)$.

   The directional derivative $D_{\mathbf{v}} h = \nabla h \cdot \mathbf{v}$.

   $$D_{\mathbf{v}} h = (e, e, e) \cdot (1, -1, 0) = e - e = 0$$

   Therefore, the derivative is 0.

3. For $p(t, s) = \sin(u(t, s))$, find $\frac{\partial p}{\partial t}$ and $\frac{\partial p}{\partial s}$.

   Solution:
   $$\frac{\partial p}{\partial t} = \frac{d}{du}(\sin u) \cdot \frac{\partial u}{\partial t} = \cos(u(t, s)) \cdot se^{ts}$$

   Similarly,
   $$\frac{\partial p}{\partial s} = \frac{d}{du}(\sin u) \cdot \frac{\partial u}{\partial s} = \cos(u(t, s)) \cdot te^{ts}$$

   Therefore,
   $$\frac{\partial p}{\partial t} = se^{ts}\cos(e^{ts}), \quad \frac{\partial p}{\partial s} = te^{ts}\cos(e^{ts}).$$

4. Find the Jacobian of $f(x, y) = (\ln(xy), x^2 - 4y)$ at $(2, 1)$.

   Solution:
   $$J_f(x, y) = \begin{bmatrix} \frac{\partial}{\partial x}(\ln(xy)) & \frac{\partial}{\partial y}(\ln(xy)) \\ \frac{\partial}{\partial x}(x^2 - 4y) & \frac{\partial}{\partial y}(x^2 - 4y) \end{bmatrix}$$

   Calculating the partial derivatives:
   $$\frac{\partial}{\partial x}(\ln(xy)) = \frac{y}{xy}, \quad \frac{\partial}{\partial y}(\ln(xy)) = \frac{x}{xy}$$

   $$\frac{\partial}{\partial x}(x^2 - 4y) = 2x, \quad \frac{\partial}{\partial y}(x^2 - 4y) = -4$$

   At $(2, 1)$:
   $$J_f(2, 1) = \begin{bmatrix} \frac{1}{2} & 1 \\ 4 & -4 \end{bmatrix}$$

   Therefore, the Jacobian is $\begin{bmatrix} \frac{1}{2} & 1 \\ 4 & -4 \end{bmatrix}$.

5. Compute the derivative of $h(x, y) = g(f(x, y)) = \ln(x^y)$.

   Solution:
   $$f(x, y) = x^y, \quad g(z) = \ln(z)$$

   $$h(x, y) = \ln(x^y) = y\ln(x)$$

   Using the chain rule:
   $$\frac{\partial h}{\partial x} = \frac{\partial}{\partial x}(y\ln(x)) = \frac{y}{x}$$

   $$\frac{\partial h}{\partial y} = \frac{\partial}{\partial y}(y\ln(x)) = \ln(x)$$

   Thus, $\nabla h = \left(\frac{y}{x}, \ln(x) \right)$.

117

6. For the activation $a = \sigma(wx + b)$, compute $\frac{da}{dw}$.

Solution:

$$a = \sigma(z), \quad z = wx + b$$

$$\sigma(z) = \frac{1}{1 + e^{-z}}, \quad \sigma'(z) = \sigma(z)(1 - \sigma(z))$$

Using the chain rule:

$$\frac{da}{dw} = \sigma'(z) \cdot \frac{dz}{dw} = \sigma(z)(1 - \sigma(z)) \cdot x$$

Therefore,

$$\frac{da}{dw} = \sigma(wx + b)(1 - \sigma(wx + b))x.$$

Practice Problems 2

1. Consider the composite function $h(x, y) = \sin(x^2 + y^2)$. Use the chain rule to find the partial derivatives $\frac{\partial h}{\partial x}$ and $\frac{\partial h}{\partial y}$.

2. Evaluate the gradient of the function $h(x, y, z) = e^{xyz}$. Use the chain rule in your computation.

3. For the functions $f(u, v) = u^2 + v$ and $g(x, y) = \ln(xy)$, compute the derivative of the composition $h(x, y) = f(g(x, y), x + y)$ with respect to x.

4. Given $h(x, y) = \sqrt{x^3 + y^3}$, find the directional derivative in the direction of vector $\mathbf{a} = \langle 3, 4 \rangle$.

5. For the neural network with layers $f_1(x) = \sigma(Wx + b)$ and $f_2(y) = \sigma(Vy + c)$, where $\sigma(z) = \frac{1}{1+e^{-z}}$, compute the derivative $\frac{\partial h}{\partial x}$ of the composite function $h(x) = f_2(f_1(x))$.

6. Consider the function $h(x, y) = \cos(x) \ln(y)$. Compute the Hessian matrix at a given point using multivariable chain rule techniques.

Answers 2

1. Consider the composite function $h(x, y) = \sin(x^2 + y^2)$. Use the chain rule to find the partial derivatives $\frac{\partial h}{\partial x}$ and $\frac{\partial h}{\partial y}$.

   Solution:

   $$\frac{\partial h}{\partial x} = \frac{d}{d(x^2 + y^2)} \left(\sin(x^2 + y^2) \right) \cdot \frac{\partial}{\partial x}(x^2 + y^2)$$
   $$= \cos(x^2 + y^2) \cdot 2x = 2x \cos(x^2 + y^2)$$

   $$\frac{\partial h}{\partial y} = \frac{d}{d(x^2 + y^2)} \left(\sin(x^2 + y^2) \right) \cdot \frac{\partial}{\partial y}(x^2 + y^2)$$
   $$= \cos(x^2 + y^2) \cdot 2y = 2y \cos(x^2 + y^2)$$

   Therefore,

   $$\frac{\partial h}{\partial x} = 2x \cos(x^2 + y^2), \quad \frac{\partial h}{\partial y} = 2y \cos(x^2 + y^2).$$

2. Evaluate the gradient of the function $h(x, y, z) = e^{xyz}$. Use the chain rule in your computation.

   Solution:

   $$\frac{\partial h}{\partial x} = e^{xyz} \cdot \frac{\partial}{\partial x}(xyz) = e^{xyz} \cdot yz$$

   $$\frac{\partial h}{\partial y} = e^{xyz} \cdot \frac{\partial}{\partial y}(xyz) = e^{xyz} \cdot xz$$

   $$\frac{\partial h}{\partial z} = e^{xyz} \cdot \frac{\partial}{\partial z}(xyz) = e^{xyz} \cdot xy$$

   Therefore, the gradient is:

   $$\nabla h = \langle yze^{xyz}, xze^{xyz}, xye^{xyz} \rangle$$

3. For the functions $f(u, v) = u^2 + v$ and $g(x, y) = \ln(xy)$, compute the derivative of the composition $h(x, y) = f(g(x, y), x + y)$ with respect to x.

   Solution: Let $u = g(x, y) = \ln(xy)$ and $v = x + y$. Then, $h(x, y) = f(u, v) = u^2 + v$.

   The derivative is:

   $$\frac{\partial h}{\partial x} = \frac{\partial f}{\partial u} \cdot \frac{\partial u}{\partial x} + \frac{\partial f}{\partial v} \cdot \frac{\partial v}{\partial x}$$

   $$\frac{\partial f}{\partial u} = 2u, \quad \frac{\partial u}{\partial x} = \frac{1}{x}$$

   $$\frac{\partial f}{\partial v} = 1, \quad \frac{\partial v}{\partial x} = 1$$

   Therefore,

   $$\frac{\partial h}{\partial x} = 2\ln(xy) \cdot \frac{1}{x} + 1 = \frac{2\ln(xy)}{x} + 1$$

4. Given $h(x, y) = \sqrt{x^3 + y^3}$, find the directional derivative in the direction of vector $\mathbf{a} = \langle 3, 4 \rangle$.

   Solution: First, normalize the vector $\mathbf{a} = \langle 3, 4 \rangle$:

   $$\|\mathbf{a}\| = \sqrt{3^2 + 4^2} = \sqrt{25} = 5$$

   $$\hat{\mathbf{a}} = \left\langle \frac{3}{5}, \frac{4}{5} \right\rangle$$

   Compute the partial derivatives:

   $$\frac{\partial h}{\partial x} = \frac{1}{2\sqrt{x^3 + y^3}} \cdot 3x^2 = \frac{3x^2}{2\sqrt{x^3 + y^3}}$$

   $$\frac{\partial h}{\partial y} = \frac{1}{2\sqrt{x^3 + y^3}} \cdot 3y^2 = \frac{3y^2}{2\sqrt{x^3 + y^3}}$$

   The directional derivative is:

   $$D_{\mathbf{a}} h = \nabla h \cdot \hat{\mathbf{a}} = \left\langle \frac{3x^2}{2\sqrt{x^3 + y^3}}, \frac{3y^2}{2\sqrt{x^3 + y^3}} \right\rangle \cdot \left\langle \frac{3}{5}, \frac{4}{5} \right\rangle$$

   $$= \frac{3x^2}{2\sqrt{x^3 + y^3}} \cdot \frac{3}{5} + \frac{3y^2}{2\sqrt{x^3 + y^3}} \cdot \frac{4}{5}$$

   $$= \frac{9x^2 + 12y^2}{10\sqrt{x^3 + y^3}}$$

120

5. For the neural network with layers $f_1(x) = \sigma(Wx + b)$ and $f_2(y) = \sigma(Vy + c)$, where $\sigma(z) = \frac{1}{1+e^{-z}}$, compute the derivative $\frac{\partial h}{\partial x}$ of the composite function $h(x) = f_2(f_1(x))$.

Solution: Using the chain rule, the derivative of the sigmoid function is:

$$\sigma'(z) = \sigma(z)(1 - \sigma(z))$$

Let $y = f_1(x) = \sigma(Wx + b)$ and $z = f_2(y) = \sigma(Vy + c)$. The derivative is:

$$\frac{\partial h}{\partial x} = \frac{\partial f_2}{\partial y} \cdot \frac{\partial f_1}{\partial x}$$

$$\frac{\partial f_2}{\partial y} = \sigma(Vy + c)(1 - \sigma(Vy + c)) \cdot V$$

$$\frac{\partial f_1}{\partial x} = \sigma(Wx + b)(1 - \sigma(Wx + b)) \cdot W$$

Therefore,

$$\frac{\partial h}{\partial x} = \sigma(Vy + c)(1 - \sigma(Vy + c)) \cdot V \cdot \sigma(Wx + b)(1 - \sigma(Wx + b)) \cdot W$$

6. Consider the function $h(x, y) = \cos(x) \ln(y)$. Compute the Hessian matrix at a given point using multivariable chain rule techniques.

Solution: First, compute the second partial derivatives:

$$\frac{\partial h}{\partial x} = -\sin(x) \ln(y)$$

$$\frac{\partial^2 h}{\partial x^2} = -\cos(x) \ln(y)$$

$$\frac{\partial h}{\partial y} = \frac{\cos(x)}{y}$$

$$\frac{\partial^2 h}{\partial y^2} = -\frac{\cos(x)}{y^2}$$

$$\frac{\partial^2 h}{\partial x \partial y} = \frac{-\sin(x)}{y}$$

The Hessian matrix is:

$$H = \begin{bmatrix} \frac{\partial^2 h}{\partial x^2} & \frac{\partial^2 h}{\partial x \partial y} \\ \frac{\partial^2 h}{\partial y \partial x} & \frac{\partial^2 h}{\partial y^2} \end{bmatrix} = \begin{bmatrix} -\cos(x) \ln(y) & \frac{-\sin(x)}{y} \\ \frac{-\sin(x)}{y} & -\frac{\cos(x)}{y^2} \end{bmatrix}$$

Therefore, the Hessian matrix is:

$$\begin{bmatrix} -\cos(x) \ln(y) & -\frac{\sin(x)}{y} \\ -\frac{\sin(x)}{y} & -\frac{\cos(x)}{y^2} \end{bmatrix}$$

Practice Problems 3

1. Prove the chain rule in multivariable calculus using the concept of directional derivatives.

2. Let $f(x, y, z) = e^{xy} \cdot \sin(z)$, and $g(t) = (t, t^2, \ln(t))$. Use the chain rule to find $\frac{d}{dt}(f \circ g)(t)$.

3. Consider the functions $f(u, v) = u^3 + v^2$ and $g(x, y) = (xy, x + y)$. Calculate the derivative of the composition $f(g(x, y))$.

4. Use the chain rule to find the gradient of the function $h(x, y, z) = \sin(xyz) + e^{x+y^2}$ at the point (1, 0, -1).

5. Verify that the multivariable chain rule holds for the functions $f(u, v) = u^2 + 3v$ and $g(x, y) = (x - y, x + y)$ by explicitly computing the derivatives and their compositions.

6. Apply the chain rule in the context of backpropagation in a neural network to show how the gradient of a loss function propagates through a sigmoid activation function.

Answers 3

1. **Solution:** The chain rule in multivariable calculus states that for $h = g(f(\mathbf{x}))$, the gradient ∇h can be expressed as $\nabla g \cdot J_f(\mathbf{x})$.

   Using directional derivatives, for any direction $\mathbf{d}$:

   $$D_{\mathbf{d}}h = \nabla g(f(\mathbf{x})) \cdot (J_f(\mathbf{x}) \cdot \mathbf{d}).$$

   This is equivalent to:
   $$\nabla h(\mathbf{x}) \cdot \mathbf{d}.$$

   By matching this with the definition of directional derivatives, we confirm the chain rule for all directions.

2. **Solution:** We start with $f(x, y, z) = e^{xy}\sin(z)$, $g(t) = (t, t^2, \ln(t))$, and want to find $\frac{d}{dt}(f \circ g)(t)$.

   Derivative of $e^{xy}\sin(z)$ at $(x, y, z) = (t, t^2, \ln(t))$:

   $$\frac{\partial f}{\partial x} = ye^{xy}\sin(z), \quad \frac{\partial f}{\partial y} = xe^{xy}\sin(z), \quad \frac{\partial f}{\partial z} = e^{xy}\cos(z).$$

   Then compute:

   $$\frac{d}{dt}(f \circ g)(t) = \frac{\partial f}{\partial x} \cdot \frac{dx}{dt} + \frac{\partial f}{\partial y} \cdot \frac{dy}{dt} + \frac{\partial f}{\partial z} \cdot \frac{dz}{dt}$$

   $$= (t^2 e^{t^3}\sin(\ln t)) \cdot 1 + (te^{t^3}\sin(\ln t)) \cdot 2t + (e^{t^3}\cos(\ln t)) \cdot \frac{1}{t}$$

   $$= t^2 e^{t^3}\sin(\ln t) + 2t^2 e^{t^3}\sin(\ln t) + \frac{e^{t^3}\cos(\ln t)}{t}$$

   $$= (t^2 + 2t^2)e^{t^3}\sin(\ln t) + \frac{e^{t^3}\cos(\ln t)}{t}$$

3. **Solution:** For $f(u, v) = u^3 + v^2$ and $g(x, y) = (xy, x + y)$, find the derivative of $f(g(x, y))$.

   Compute the Jacobians:

   $$J_g(x, y) = \begin{bmatrix} y & x \\ 1 & 1 \end{bmatrix}, \quad \text{at } (x, y)$$

   $$\frac{\partial f}{\partial u} = 3u^2, \quad \frac{\partial f}{\partial v} = 2v$$

   Then:

   $$J_f(u, v) = \begin{bmatrix} 3u^2 & 2v \end{bmatrix}$$

123

Using the chain rule:

$$J_f(g) \cdot J_g(x, y) = \begin{bmatrix} 3(xy)^2 & 2(x + y) \end{bmatrix} \cdot \begin{bmatrix} y & x \\ 1 & 1 \end{bmatrix}$$

$$= \begin{bmatrix} 3(xy)^2 \cdot y + 2(x + y), 3(xy)^2 \cdot x + 2(x + y) \end{bmatrix}$$

4. **Solution:** To find the gradient of $h(x, y, z) = \sin(xyz) + e^{x+y^2}$ at (1, 0, -1), compute:

$$\frac{\partial h}{\partial x} = yz \cos(xyz) + e^{x+y^2}$$

$$\frac{\partial h}{\partial y} = xz \cos(xyz) + 2ye^{x+y^2}$$

$$\frac{\partial h}{\partial z} = xy \cos(xyz)$$

Evaluate at (1, 0, -1):

$$\frac{\partial h}{\partial x} = 0 \cdot (-1) \cos(0) + e^1 = e$$

$$\frac{\partial h}{\partial y} = 1 \cdot (-1) \cos(0) + 2 \cdot 0 \cdot e^1 = -1$$

$$\frac{\partial h}{\partial z} = 1 \cdot 0 \cos(0) = 0$$

Gradient:

$$\nabla h(1, 0, -1) = (e, -1, 0)$$

5. **Solution:** Verify chain rule for $f(u, v) = u^2 + 3v$ and $g(x, y) = (x - y, x + y)$.

Calculate:

$$J_g(x, y) = \begin{bmatrix} 1 & -1 \\ 1 & 1 \end{bmatrix}$$

$$\frac{\partial f}{\partial u} = 2u, \quad \frac{\partial f}{\partial v} = 3$$

$$J_f(u, v) = \begin{bmatrix} 2u & 3 \end{bmatrix}$$

Apply chain rule:

$$J_h(x, y) = J_f(g) \cdot J_g(x, y) = \begin{bmatrix} 2(x - y) & 3 \end{bmatrix} \cdot \begin{bmatrix} 1 & -1 \\ 1 & 1 \end{bmatrix}$$

$$= \begin{bmatrix} 2(x - y) \cdot 1 + 3, 2(x - y) \cdot (-1) + 3 \end{bmatrix}$$

Verify manually:

$$h(x, y) = (x - y)^2 + 3(x + y)$$

6. **Solution:** In backpropagation, for a sigmoid activation function $\sigma(x) = \frac{1}{1+e^{-x}}$, the gradient of the loss with respect to the parameters using the chain rule is:

$$\sigma'(x) = \sigma(x)(1 - \sigma(x))$$

Suppose the loss is $L = \frac{1}{2}(y - \hat{y})^2$ where $\hat{y} = \sigma(w \cdot x + b)$. Compute the derivative using the chain rule:

$$\frac{dL}{dw} = \frac{dL}{d\hat{y}} \cdot \frac{d\hat{y}}{dz} \cdot \frac{dz}{dw}$$

124

$$\frac{dL}{d\hat{y}} = -(y - \hat{y}), \quad \frac{d\hat{y}}{dz} = \sigma(z)(1 - \sigma(z)), \quad \frac{dz}{dw} = x$$

Combine terms:

$$\frac{dL}{dw} = -(y - \hat{y}) \cdot \sigma(z)(1 - \sigma(z)) \cdot x$$

Chapter 13

Integration Basics

.

Practice Problems 1

1. Evaluate the indefinite integral:

$$\int (3x^2 + 4x + 1)\, dx$$

2. Find the integral of the exponential function:

$$\int e^{3x}\, dx$$

3. Determine the indefinite integral using substitution method:

$$\int x\sqrt{x+1}\, dx$$

4. Compute the definite integral:

$$\int_1^4 (2x^3 - x^2)\, dx$$

5. Solve the integration problem for a trigonometric function:

$$\int \sin(2x)\, dx$$

6. Calculate the integral of a logarithmic function:

$$\int x \ln(x)\, dx$$

Answers 1

1. Evaluate the indefinite integral:

$$\int (3x^2 + 4x + 1)\, dx$$

   Solution:

$$\int (3x^2 + 4x + 1)\, dx = \int 3x^2\, dx + \int 4x\, dx + \int 1\, dx$$

$$= \frac{3x^{2+1}}{2+1} + \frac{4x^{1+1}}{1+1} + x + C$$

$$= x^3 + 2x^2 + x + C$$

Therefore,

$$\int (3x^2 + 4x + 1)\, dx = x^3 + 2x^2 + x + C.$$

2. Find the integral of the exponential function:

$$\int e^{3x}\, dx$$

Solution:

$$\int e^{3x}\, dx = \frac{1}{3}e^{3x} + C$$

This follows from the rule $\int e^{ax}\, dx = \frac{1}{a}e^{ax} + C$. Therefore,

$$\int e^{3x}\, dx = \frac{1}{3}e^{3x} + C.$$

3. Determine the indefinite integral using substitution method:

$$\int x\sqrt{x+1}\, dx$$

Solution: Let $u = x + 1$, then $du = dx$ and $x = u - 1$.

$$\int x\sqrt{x+1}\, dx = \int (u-1)\sqrt{u}\, du$$

$$= \int (u^{3/2} - u^{1/2})\, du = \int u^{3/2}\, du - \int u^{1/2}\, du$$

$$= \frac{2}{5}u^{5/2} - \frac{2}{3}u^{3/2} + C$$

Substitute $u = x + 1$ back:

$$= \frac{2}{5}(x+1)^{5/2} - \frac{2}{3}(x+1)^{3/2} + C$$

4. Compute the definite integral:

$$\int_1^4 (2x^3 - x^2)\, dx$$

Solution: First find the indefinite integral:

$$\int (2x^3 - x^2)\, dx = \frac{2x^{3+1}}{3+1} - \frac{x^{2+1}}{2+1} + C$$

$$= \frac{1}{2}x^4 - \frac{1}{3}x^3 + C$$

Evaluate the definite integral from 1 to 4:

$$= \left[\frac{1}{2}(4)^4 - \frac{1}{3}(4)^3\right] - \left[\frac{1}{2}(1)^4 - \frac{1}{3}(1)^3\right]$$

$$= \left[\frac{1}{2}(256) - \frac{1}{3}(64)\right] - \left[\frac{1}{2} - \frac{1}{3}\right]$$

128

$$= [128 - \frac{64}{3}] - [\frac{3}{6} - \frac{2}{6}]$$

$$= \frac{384}{3} - \frac{64}{3} - \frac{1}{6}$$

$$= \frac{320}{3} - \frac{1}{6} = \frac{640 - 1}{6} = \frac{639}{6} = \frac{213}{2}$$

5. Solve the integration problem for a trigonometric function:

$$\int \sin(2x)\, dx$$

Solution: Using the substitution $u = 2x$, $du = 2\, dx$, thus $dx = \frac{1}{2} du$:

$$\int \sin(2x)\, dx = \int \sin(u) \cdot \frac{1}{2}\, du$$

$$= \frac{1}{2} \int \sin(u)\, du = -\frac{1}{2} \cos(u) + C$$

Substitute $u = 2x$ back:

$$= -\frac{1}{2} \cos(2x) + C$$

6. Calculate the integral of a logarithmic function:

$$\int x \ln(x)\, dx$$

Solution: Use integration by parts, where $u = \ln(x)$ and $dv = x\, dx$. Then $du = \frac{1}{x}\, dx$ and $v = \frac{x^2}{2}$:

$$\int x \ln(x)\, dx = \ln(x) \cdot \frac{x^2}{2} - \int \frac{x^2}{2} \cdot \frac{1}{x}\, dx$$

$$= \frac{x^2}{2} \ln(x) - \frac{1}{2} \int x\, dx$$

$$= \frac{x^2}{2} \ln(x) - \frac{1}{2} \cdot \frac{x^2}{2} + C$$

$$= \frac{x^2}{2} \ln(x) - \frac{x^2}{4} + C$$

Practice Problems 2

1. Evaluate the indefinite integral:

$$\int 3x^2\, dx$$

2. Evaluate the indefinite integral using substitution:

$$\int 2xe^{x^2}\, dx$$

3. Find the integral of the following trigonometric function:

$$\int \sin(2x)\, dx$$

4. Determine the area under the curve of the following function over the interval $[0,\ 1]$:

$$\int_0^1 (4 - x^2)\, dx$$

5. Calculate the integral using integration by parts:

$$\int x \ln(x)\, dx$$

6. Evaluate the expected value of a random variable with probability density function given by $f(x) = 3x^2$ over the interval $[0, 1]$:

$$\mathbb{E}[X] = \int_0^1 x \cdot 3x^2 \, dx$$

Answers 2

1. Evaluate the indefinite integral:

$$\int 3x^2 \, dx$$

Solution: Using the power rule for integration, $\int x^n \, dx = \frac{x^{n+1}}{n+1} + C$, where $n = 2$,

$$\int 3x^2 \, dx = 3 \cdot \frac{x^{2+1}}{2+1} + C = \frac{3x^3}{3} + C = x^3 + C$$

Therefore,

$$\int 3x^2 \, dx = x^3 + C.$$

2. Evaluate the indefinite integral using substitution:

$$\int 2xe^{x^2} \, dx$$

Solution: Use substitution: let $u = x^2$ then $du = 2x \, dx$. Therefore,

$$\int 2xe^{x^2} \, dx = \int e^u \, du = e^u + C = e^{x^2} + C.$$

Therefore,

$$\int 2xe^{x^2} \, dx = e^{x^2} + C.$$

3. Find the integral of the following trigonometric function:

$$\int \sin(2x) \, dx$$

Solution: Using substitution, let $u = 2x$ then $du = 2 \, dx$ or $dx = \frac{du}{2}$,

$$\int \sin(2x) \, dx = \frac{1}{2} \int \sin(u) \, du = -\frac{1}{2} \cos(u) + C = -\frac{1}{2} \cos(2x) + C.$$

Therefore,

$$\int \sin(2x) \, dx = -\frac{1}{2} \cos(2x) + C.$$

4. Determine the area under the curve of the following function over the interval $[0, 1]$:

$$\int_0^1 (4 - x^2) \, dx$$

Solution: Integrate term by term:

$$\int_0^1 (4 - x^2) \, dx = \left[4x - \frac{x^3}{3} \right]_0^1 = \left(4(1) - \frac{1^3}{3} \right) - \left(4(0) - \frac{0^3}{3} \right)$$

$$= \left(4 - \frac{1}{3} \right) = \frac{12}{3} - \frac{1}{3} = \frac{11}{3}.$$

Therefore, the area is

$$\frac{11}{3}.$$

5. Calculate the integral using integration by parts:

$$\int x \ln(x) \, dx$$

Solution: Let $u = \ln(x)$ and $dv = x \, dx$. Then $du = \frac{1}{x} \, dx$ and $v = \frac{x^2}{2}$.

$$\int x \ln(x) \, dx = \frac{x^2}{2} \ln(x) - \int \frac{x^2}{2} \cdot \frac{1}{x} \, dx = \frac{x^2}{2} \ln(x) - \frac{1}{2} \int x \, dx$$

$$= \frac{x^2}{2} \ln(x) - \frac{1}{2} \left(\frac{x^2}{2} \right) + C = \frac{x^2}{2} \ln(x) - \frac{x^2}{4} + C.$$

Therefore,

$$\int x \ln(x) \, dx = \frac{x^2}{2} \ln(x) - \frac{x^2}{4} + C.$$

6. Evaluate the expected value of a random variable with probability density function given by $f(x) = 3x^2$ over the interval $[0, 1]$:

$$\mathbb{E}[X] = \int_0^1 x \cdot 3x^2 \, dx$$

Solution:

$$\mathbb{E}[X] = \int_0^1 3x^3 \, dx = \left[\frac{3x^4}{4} \right]_0^1 = \frac{3(1)^4}{4} - \frac{3(0)^4}{4}$$

$$= \frac{3}{4}.$$

Therefore, the expected value is

$$\frac{3}{4}.$$

Practice Problems 3

1. Evaluate the integral of the function using the power rule:

$$\int (5x^3 - 4x + 1) \, dx$$

2. Compute the integral of the exponential function:

$$\int e^{3x} \, dx$$

3. Determine the antiderivative of the logarithmic function:

$$\int \ln(x) \, dx$$

4. Find the integral using a basic trigonometric formula:

$$\int \sin(2x) \, dx$$

5. Calculate the expected value for a given probability density function:

$$f(x) = 2x, \quad x \in [0, 1]$$

$$\mathbb{E}[X] = \int_0^1 x f(x) \, dx$$

6. Use the Trapezoidal Rule to approximate the integral of $f(x) = x^2$ over the interval $[0, 2]$ using 4 subintervals.

Answers 3

1. Evaluate the integral of the function using the power rule:

$$\int (5x^3 - 4x + 1)\, dx$$

Solution:

$$\int 5x^3\, dx = \frac{5}{4}x^4 + C_1$$

$$\int -4x\, dx = -2x^2 + C_2$$

$$\int 1\, dx = x + C_3$$

Combine the results:

$$\int (5x^3 - 4x + 1)\, dx = \frac{5}{4}x^4 - 2x^2 + x + C$$

2. Compute the integral of the exponential function:

$$\int e^{3x}\, dx$$

Solution: Using the substitution $u = 3x$, $du = 3\, dx$, so $dx = \frac{1}{3}\, du$:

$$\int e^{3x}\, dx = \frac{1}{3}\int e^u\, du = \frac{1}{3}e^u + C = \frac{1}{3}e^{3x} + C$$

3. Determine the antiderivative of the logarithmic function:

$$\int \ln(x)\, dx$$

Solution: Use integration by parts: Let $u = \ln(x)$, $dv = dx$. Therefore, $du = \frac{1}{x}\, dx$, $v = x$.

$$\int \ln(x)\, dx = x\ln(x) - \int x \cdot \frac{1}{x}\, dx$$

$$= x\ln(x) - \int 1\, dx = x\ln(x) - x + C$$

4. Find the integral using a basic trigonometric formula:

$$\int \sin(2x)\,dx$$

Solution: Use the substitution $u = 2x$, $du = 2\,dx$, so $dx = \frac{1}{2}\,du$:

$$\int \sin(2x)\,dx = \frac{1}{2} \int \sin(u)\,du = -\frac{1}{2}\cos(u) + C$$

$$= -\frac{1}{2}\cos(2x) + C$$

5. Calculate the expected value for a given probability density function:

$$f(x) = 2x, \quad x \in [0, 1]$$

$$\mathbb{E}[X] = \int_0^1 x f(x)\,dx$$

Solution: Substitute $f(x) = 2x$:

$$\int_0^1 x(2x)\,dx = 2 \int_0^1 x^2\,dx$$

$$= 2 \left[\frac{x^3}{3}\right]_0^1 = 2 \left(\frac{1^3}{3} - \frac{0^3}{3}\right)$$

$$= \frac{2}{3}$$

6. Use the Trapezoidal Rule to approximate the integral of $f(x) = x^2$ over the interval $[0, 2]$ using 4 subintervals.
   Solution: Partition the interval $[0, 2]$ into 4 equal-length subintervals, each of width $h = \frac{2-0}{4} = 0.5$. The endpoints are $0, 0.5, 1, 1.5, 2$.

   Calculate $f(x)$ at each endpoint:

$$f(0) = 0^2 = 0, \quad f(0.5) = (0.5)^2 = 0.25, \quad f(1) = 1^2 = 1$$

$$f(1.5) = (1.5)^2 = 2.25, \quad f(2) = 2^2 = 4$$

   Apply the Trapezoidal Rule:

$$\text{Approximation} = \frac{h}{2}\left(f(0) + 2f(0.5) + 2f(1) + 2f(1.5) + f(2)\right)$$

$$= \frac{0.5}{2}\left(0 + 2(0.25) + 2(1) + 2(2.25) + 4\right)$$

$$= 0.25\left(0 + 0.5 + 2 + 4.5 + 4\right) = 0.25 \times 11 = 2.75$$

135

Chapter 14

Techniques of Integration

Practice Problems 1

1. Evaluate the following integral using integration by parts:

$$\int x \ln(x)\, dx$$

2. Solve the integral using trigonometric integrals:

$$\int \sin^3(x)\, dx$$

3. Perform trigonometric substitution to evaluate the following integral:

$$\int \frac{1}{\sqrt{9 - x^2}}\, dx$$

4. Decompose and evaluate the given rational function using partial fraction decomposition:

$$\int \frac{3x + 5}{(x - 1)(x + 2)}\, dx$$

5. Evaluate the following integral using trigonometric substitution:

$$\int x^2 \sqrt{x^2 + 4}\, dx$$

6. Use the technique of trigonometric integrals to solve:

$$\int \cos^2(x) \sin(x)\, dx$$

Answers 1

1. Evaluate the following integral using integration by parts:

$$\int x \ln(x)\, dx$$

Solution: Let $u = \ln(x)$ and $dv = x\, dx$. Then $du = \frac{1}{x}\, dx$ and $v = \frac{x^2}{2}$. Applying the integration by parts formula:

$$\int u\, dv = uv - \int v\, du$$

$$= \frac{x^2}{2} \ln(x) - \int \frac{x^2}{2} \cdot \frac{1}{x} \, dx$$

$$= \frac{x^2}{2} \ln(x) - \int \frac{x}{2} \, dx$$

$$= \frac{x^2}{2} \ln(x) - \frac{x^2}{4} + C$$

2. Solve the integral using trigonometric integrals:

$$\int \sin^3(x) \, dx$$

Solution: Rewrite $\sin^3(x)$ as $\sin(x) \cdot \sin^2(x)$ and use the identity $\sin^2(x) = 1 - \cos^2(x)$:

$$= \int \sin(x)(1 - \cos^2(x)) \, dx$$

$$= \int \sin(x) \, dx - \int \sin(x) \cos^2(x) \, dx$$

The first integral is:

$$= -\cos(x)$$

For the second integral, use substitution $u = \cos(x)$, $du = -\sin(x) \, dx$:

$$\int u^2 \, du = \frac{u^3}{3} = \frac{\cos^3(x)}{3}$$

So,

$$= -\cos(x) + \frac{\cos^3(x)}{3} + C$$

3. Perform trigonometric substitution to evaluate the following integral:

$$\int \frac{1}{\sqrt{9 - x^2}} \, dx$$

Solution: Use the substitution $x = 3\sin(\theta)$, where $dx = 3\cos(\theta) \, d\theta$ and $\sqrt{9 - x^2} = 3\cos(\theta)$:

$$\int \frac{1}{3\cos(\theta)} \cdot 3\cos(\theta) \, d\theta = \int 1 \, d\theta = \theta + C$$

Since $x = 3\sin(\theta)$, then $\theta = \arcsin(x/3)$:

$$= \arcsin(x/3) + C$$

4. Decompose and evaluate the given rational function using partial fraction decomposition:

$$\int \frac{3x + 5}{(x - 1)(x + 2)} \, dx$$

Solution: Decompose as:

$$\frac{3x + 5}{(x - 1)(x + 2)} = \frac{A}{x - 1} + \frac{B}{x + 2}$$

Solving for A and B, we have:

$$3x + 5 = A(x + 2) + B(x - 1)$$

Setting $x = 1$, we find $B = 2$. Setting $x = -2$, we find $A = 1$. So:

$$\int \left(\frac{1}{x - 1} + \frac{2}{x + 2} \right) \, dx = \int \frac{1}{x - 1} \, dx + \int \frac{2}{x + 2} \, dx$$

$$= \ln|x - 1| + 2\ln|x + 2| + C$$

138

5. Evaluate the following integral using trigonometric substitution:

$$\int x^2 \sqrt{x^2 + 4} \, dx$$

Solution: Use the substitution $x = 2\tan(\theta)$, $dx = 2\sec^2(\theta) \, d\theta$ and $\sqrt{x^2 + 4} = 2\sec(\theta)$:

$$= \int (2\tan(\theta))^2 \cdot 2\sec(\theta) \cdot 2\sec^2(\theta) \, d\theta$$

$$= 8 \int \tan^2(\theta) \sec^3(\theta) \, d\theta$$

Write $\tan^2(\theta) = \sec^2(\theta) - 1$:

$$= 8 \int (\sec^5(\theta) - \sec^3(\theta)) \, d\theta$$

Integrate using known formulas or reduction:

$$= \text{(long integral evaluation with respect to theta)}$$

Simplifying, back-substitute $x = 2\tan(\theta)$ to express in terms of x.

6. Use the technique of trigonometric integrals to solve:

$$\int \cos^2(x) \sin(x) \, dx$$

Solution: Use substitution $u = \cos(x)$, $du = -\sin(x) \, dx$:

$$= -\int u^2 \, du = -\frac{u^3}{3} + C = -\frac{\cos^3(x)}{3} + C$$

Practice Problems 2

1. Evaluate the following integral using integration by parts:

$$\int x \ln(x) \, dx$$

2. Solve the integral using trigonometric substitution:

$$\int \frac{dx}{\sqrt{9 - x^2}}$$

139

3. Evaluate the integral using partial fraction decomposition:

$$\int \frac{3x+5}{x^2-4x+3}\,dx$$

4. Solve the integral using trigonometric identities:

$$\int \sin^3(x)\cos^2(x)\,dx$$

5. Evaluate the integral using the technique of trigonometric substitution:

$$\int \sqrt{x^2+1}\,dx$$

6. Use integration by parts to find the integral:

$$\int e^x \cos(x)\,dx$$

Answers 2

1. Evaluate the following integral using integration by parts:

$$\int x \ln(x) \, dx$$

Solution: Let $u = \ln(x)$ and $dv = x \, dx$. Then $du = \frac{1}{x} \, dx$ and $v = \frac{x^2}{2}$.

$$\int x \ln(x) \, dx = \frac{x^2}{2} \ln(x) - \int \frac{x^2}{2} \cdot \frac{1}{x} \, dx$$

$$= \frac{x^2}{2} \ln(x) - \frac{1}{2} \int x \, dx$$

$$= \frac{x^2}{2} \ln(x) - \frac{1}{4} x^2 + C$$

Therefore,

$$\int x \ln(x) \, dx = \frac{x^2}{2} \ln(x) - \frac{1}{4} x^2 + C.$$

2. Solve the integral using trigonometric substitution:

$$\int \frac{dx}{\sqrt{9 - x^2}}$$

Solution: Use the substitution $x = 3\sin(\theta)$, then $dx = 3\cos(\theta) \, d\theta$ and $\sqrt{9 - x^2} = 3\cos(\theta)$.

$$\int \frac{dx}{\sqrt{9 - x^2}} = \int \frac{3\cos(\theta) \, d\theta}{3\cos(\theta)}$$

$$= \int d\theta = \theta + C$$

Since $\theta = \arcsin(\frac{x}{3})$,

$$\int \frac{dx}{\sqrt{9 - x^2}} = \arcsin\left(\frac{x}{3}\right) + C.$$

3. Evaluate the integral using partial fraction decomposition:

$$\int \frac{3x + 5}{x^2 - 4x + 3} \, dx$$

Solution: Factor the denominator: $x^2 - 4x + 3 = (x - 3)(x - 1)$.

$$\frac{3x + 5}{x^2 - 4x + 3} = \frac{A}{x - 3} + \frac{B}{x - 1}$$

Solving for A and B,

$$3x + 5 = A(x - 1) + B(x - 3)$$

By equating coefficients, $A = 2$, $B = 1$.

$$\int \left(\frac{2}{x - 3} + \frac{1}{x - 1} \right) dx = 2\ln|x - 3| + \ln|x - 1| + C$$

Therefore,

$$\int \frac{3x + 5}{x^2 - 4x + 3} \, dx = 2\ln|x - 3| + \ln|x - 1| + C.$$

4. Solve the integral using trigonometric identities:

$$\int \sin^3(x) \cos^2(x) \, dx$$

Solution: Use $\sin^2(x) = 1 - \cos^2(x)$.

$$\int \sin(x)(1 - \cos^2(x)) \cos^2(x) \, dx = \int \sin(x)(\cos^2(x) - \cos^4(x)) \, dx$$

Let $u = \cos(x)$, $du = -\sin(x) \, dx$.

$$\int (u^2 - u^4)(-du) = \int (u^4 - u^2) \, du$$

$$= \frac{u^5}{5} - \frac{u^3}{3} + C$$

Substitute $u = \cos(x)$,

$$= \frac{\cos^5(x)}{5} - \frac{\cos^3(x)}{3} + C.$$

5. Evaluate the integral using trigonometric substitution:

$$\int \sqrt{x^2 + 1} \, dx$$

Solution: Use the substitution $x = \sinh(\theta)$, then $dx = \cosh(\theta) \, d\theta$ and $\sqrt{x^2 + 1} = \cosh(\theta)$.

$$\int \cosh^2(\theta) \, d\theta$$

Use $\cosh^2(\theta) = \frac{1 + \cosh(2\theta)}{2}$,

$$\int \frac{1 + \cosh(2\theta)}{2} \, d\theta = \frac{1}{2} \left[\theta + \frac{1}{2} \sinh(2\theta) \right] + C$$

Back-substitute using $\theta = \ln(x + \sqrt{x^2 + 1})$,

$$= \frac{1}{2}[\ln(x + \sqrt{x^2 + 1})] + \frac{1}{4}(x\sqrt{x^2 + 1} + \ln(x + \sqrt{x^2 + 1}))]$$

Therefore,

$$\int \sqrt{x^2 + 1} \, dx = \frac{1}{2}x\sqrt{x^2 + 1} + \frac{1}{2} \ln(x + \sqrt{x^2 + 1}) + C.$$

6. Use integration by parts to find the integral:

$$\int e^x \cos(x) \, dx$$

Solution: Let $u = \cos(x)$, $dv = e^x \, dx$, then $du = -\sin(x) \, dx$, $v = e^x$.

$$\int e^x \cos(x) \, dx = e^x \cos(x) + \int e^x \sin(x) \, dx$$

Apply integration by parts to $\int e^x \sin(x) \, dx$, letting $u = \sin(x)$, $dv = e^x \, dx$.

$$\int e^x \sin(x) \, dx = e^x \sin(x) - \int e^x \cos(x) \, dx$$

Thus,

$$I = e^x \cos(x) + (e^x \sin(x) - I)$$
$$2I = e^x(\cos(x) + \sin(x))$$
$$I = \frac{e^x(\cos(x) + \sin(x))}{2} + C$$

Therefore,

$$\int e^x \cos(x)\, dx = \frac{e^x(\cos(x) + \sin(x))}{2} + C.$$

Practice Problems 3

1. Evaluate the following integral using integration by parts:

$$\int x e^x \, dx$$

2. Solve the following integral using trigonometric identities:

$$\int \sin^2(x)\, dx$$

3. Use trigonometric substitution to solve the integral:

$$\int \sqrt{1 - x^2}\, dx$$

4. Decompose the following rational function and integrate:

$$\int \frac{2x + 3}{x^2 - x - 6} \, dx$$

5. Solve the following integral:

$$\int \frac{1}{x^2 + 4} \, dx$$

using a suitable substitution method.

6. Evaluate the integral using integration by parts:

$$\int \ln(x) \, dx$$

Answers 3

1. Solve the following integral using integration by parts:

$$\int xe^x \, dx$$

Solution: Let $u = x$ and $dv = e^x \, dx$. Then $du = dx$ and $v = e^x$.

$$\int xe^x \, dx = uv - \int v \, du = xe^x - \int e^x \, dx$$

$$= xe^x - e^x + C$$

Therefore,

$$\int xe^x \, dx = xe^x - e^x + C.$$

2. Solve the following integral using trigonometric identities:

$$\int \sin^2(x) \, dx$$

Solution: Using the identity $\sin^2(x) = \frac{1-\cos(2x)}{2}$,

$$\int \sin^2(x) \, dx = \int \frac{1 - \cos(2x)}{2} \, dx$$

$$= \frac{1}{2} \left(\int 1 \, dx - \int \cos(2x) \, dx \right)$$

$$= \frac{1}{2} \left(x - \frac{1}{2} \sin(2x) \right) + C$$

Therefore,

$$\int \sin^2(x) \, dx = \frac{x}{2} - \frac{1}{4} \sin(2x) + C.$$

3. Use trigonometric substitution to solve the integral:

$$\int \sqrt{1 - x^2} \, dx$$

Solution: Use $x = \sin(\theta)$, then $dx = \cos(\theta) \, d\theta$, and $\sqrt{1 - x^2} = \cos(\theta)$.

$$\int \sqrt{1 - x^2} \, dx = \int \cos^2(\theta) \, d\theta$$

$$= \int \frac{1 + \cos(2\theta)}{2} \, d\theta = \frac{1}{2} \left(\int 1 \, d\theta + \int \cos(2\theta) \, d\theta \right)$$

$$= \frac{1}{2} \left(\theta + \frac{1}{2} \sin(2\theta) \right) + C$$

Convert back to x: $\theta = \sin^{-1}(x)$, $\sin(2\theta) = 2x\sqrt{1 - x^2}$.

$$= \frac{1}{2} \left(\sin^{-1}(x) + x\sqrt{1 - x^2} \right) + C$$

Thus,

$$\int \sqrt{1 - x^2} \, dx = \frac{1}{2} (\sin^{-1}(x) + x\sqrt{1 - x^2}) + C.$$

4. Decompose the following rational function and integrate:

$$\int \frac{2x + 3}{x^2 - x - 6} \, dx$$

Solution: First, factor the denominator: $x^2 - x - 6 = (x - 3)(x + 2)$. Decompose the fraction:

$$\frac{2x + 3}{x^2 - x - 6} = \frac{A}{x - 3} + \frac{B}{x + 2}$$

145

Solving for A and B, we have:
$$2x + 3 = A(x + 2) + B(x - 3)$$

Setting $x = 3$, $6 + 3 = A(5)$ gives $A = \frac{9}{5}$. Setting $x = -2$, $-4 + 3 = B(-5)$ gives $B = -\frac{1}{5}$. Thus decomposition gives:
$$\frac{9}{5(x - 3)} - \frac{1}{5(x + 2)}$$

Integrating each fraction:
$$\int \frac{9}{5(x - 3)} \, dx = \frac{9}{5} \ln|x - 3| + C_1$$
$$\int \frac{-1}{5(x + 2)} \, dx = -\frac{1}{5} \ln|x + 2| + C_2$$

Therefore,
$$\int \frac{2x + 3}{x^2 - x - 6} \, dx = \frac{9}{5} \ln|x - 3| - \frac{1}{5} \ln|x + 2| + C.$$

5. Solve the following integral:
$$\int \frac{1}{x^2 + 4} \, dx$$

Solution: Use the substitution $x = 2\tan(\theta)$, then $dx = 2\sec^2(\theta) \, d\theta$ and $x^2 + 4 = 4\sec^2(\theta)$.
$$\int \frac{1}{x^2 + 4} \, dx = \int \frac{2\sec^2(\theta)}{4\sec^2(\theta)} \, d\theta = \frac{1}{2} \int d\theta$$
$$= \frac{1}{2}\theta + C$$

Convert back to x: $\theta = \tan^{-1}(\frac{x}{2})$.
$$= \frac{1}{2} \tan^{-1}\left(\frac{x}{2}\right) + C$$

Therefore,
$$\int \frac{1}{x^2 + 4} \, dx = \frac{1}{2} \tan^{-1}\left(\frac{x}{2}\right) + C$$

6. Evaluate the integral using integration by parts:
$$\int \ln(x) \, dx$$

Solution: Let $u = \ln(x)$ and $dv = dx$. Then $du = \frac{1}{x}dx$ and $v = x$.
$$\int \ln(x) \, dx = uv - \int v \, du = x \ln(x) - \int x \cdot \frac{1}{x} \, dx$$
$$= x \ln(x) - \int 1 \, dx$$
$$= x \ln(x) - x + C$$

Therefore,
$$\int \ln(x) \, dx = x \ln(x) - x + C.$$

146

Chapter 15

Applications of Integration

Practice Problems 1

1. Compute the area under the curve for the function:

$$f(x) = 2x^3 - x^2 + 3$$

over the interval $[1, 2]$.

2. Find the volume of the solid obtained by revolving the curve:

$$f(x) = \sqrt{x}$$

about the x-axis over the interval $[1, 4]$, using the disk method.

3. Calculate the volume of the solid obtained by revolving the region enclosed by:

$$f(x) = x^2 \quad \text{and} \quad g(x) = x$$

about the x-axis over the interval $[0, 1]$, using the washer method.

4. Compute the expected value for the probability density function:

$$f(x) = 3x^2$$

over the interval $[0, 1]$.

5. Find the variance of the random variable with probability density function:

$$f(x) = 6x(1 - x)$$

over the interval $[0, 1]$.

6. Determine the volume of the solid obtained by revolving the curve:

$$f(y) = y^2$$

about the y-axis over the interval $[0, 2]$, using the shell method.

Answers 1

1. Compute the area under the curve for the function:

$$f(x) = 2x^3 - x^2 + 3$$

over the interval $[1, 2]$.

Solution:

$$A = \int_1^2 (2x^3 - x^2 + 3)\, dx$$

$$= \left[\frac{2}{4}x^4 - \frac{1}{3}x^3 + 3x\right]_1^2$$

$$= \left[\frac{1}{2}(2)^4 - \frac{1}{3}(2)^3 + 3(2)\right] - \left[\frac{1}{2}(1)^4 - \frac{1}{3}(1)^3 + 3(1)\right]$$

$$= \left[8 - \frac{8}{3} + 6\right] - \left[\frac{1}{2} - \frac{1}{3} + 3\right]$$

$$= \left[14 - \frac{8}{3}\right] - \left[\frac{7}{6} + 3\right]$$

$$= \left[\frac{42}{3} - \frac{8}{3}\right] - \left[\frac{43}{6}\right]$$

$$= \frac{34}{3} - \frac{43}{6}$$

$$= \frac{68}{6} - \frac{43}{6} = \frac{25}{6}$$

Therefore, the area is $\frac{25}{6}$.

2. Find the volume of the solid obtained by revolving the curve:

$$f(x) = \sqrt{x}$$

about the x-axis over the interval $[1, 4]$, using the disk method.

Solution:

$$V = \pi \int_1^4 [\sqrt{x}]^2 \, dx$$

$$= \pi \int_1^4 x \, dx$$

$$= \pi \left[\frac{x^2}{2}\right]_1^4$$

$$= \pi \left[\frac{16}{2} - \frac{1}{2}\right]$$

$$= \pi \left[\frac{15}{2}\right]$$

$$= \frac{15\pi}{2}$$

Therefore, the volume is $\frac{15\pi}{2}$.

3. Calculate the volume of the solid obtained by revolving the region enclosed by:

$$f(x) = x^2 \quad \text{and} \quad g(x) = x$$

about the x-axis over the interval $[0, 1]$, using the washer method.

Solution:

$$V = \pi \int_0^1 \left([x]^2 - [x^2]^2\right) \, dx$$

$$= \pi \int_0^1 \left(x^2 - x^4\right) \, dx$$

$$= \pi \left[\frac{x^3}{3} - \frac{x^5}{5}\right]_0^1$$

149

$$= \pi \left[\frac{1}{3} - \frac{1}{5} \right]$$

$$= \pi \left[\frac{5}{15} - \frac{3}{15} \right]$$

$$= \frac{2\pi}{15}$$

Therefore, the volume is $\frac{2\pi}{15}$.

4. Compute the expected value for the probability density function:

$$f(x) = 3x^2$$

over the interval $[0, 1]$.

Solution:

$$E[X] = \int_0^1 x \cdot 3x^2 \, dx$$

$$= 3 \int_0^1 x^3 \, dx$$

$$= 3 \left[\frac{x^4}{4} \right]_0^1$$

$$= 3 \cdot \frac{1}{4}$$

$$= \frac{3}{4}$$

Therefore, the expected value is $\frac{3}{4}$.

5. Find the variance of the random variable with probability density function:

$$f(x) = 6x(1 - x)$$

over the interval $[0, 1]$.

Solution:

$$E[X] = \int_0^1 x \cdot 6x(1 - x) \, dx$$

$$= 6 \int_0^1 x^2(1 - x) \, dx$$

$$= 6 \int_0^1 (x^2 - x^3) \, dx$$

$$= 6 \left[\frac{x^3}{3} - \frac{x^4}{4} \right]_0^1$$

$$= 6 \left[\frac{1}{3} - \frac{1}{4} \right]$$

$$= 6 \left[\frac{4}{12} - \frac{3}{12} \right]$$

$$= 6 \cdot \frac{1}{12}$$

$$= \frac{1}{2}$$

150

To compute the variance, we need $E[X^2]$:

$$E[X^2] = \int_0^1 x^2 \cdot 6x(1-x)\,dx$$

$$= 6 \int_0^1 x^3(1-x)\,dx$$

$$= 6 \int_0^1 (x^3 - x^4)\,dx$$

$$= 6 \left[\frac{x^4}{4} - \frac{x^5}{5} \right]_0^1$$

$$= 6 \left[\frac{1}{4} - \frac{1}{5} \right]$$

$$= 6 \left[\frac{5}{20} - \frac{4}{20} \right]$$

$$= 6 \cdot \frac{1}{20}$$

$$= \frac{3}{10}$$

Therefore, $\text{Var}(X) = E[X^2] - (E[X])^2 = \frac{3}{10} - \left(\frac{1}{2}\right)^2$

$$= \frac{3}{10} - \frac{1}{4}$$

$$= \frac{3}{10} - \frac{2.5}{10}$$

$$= \frac{0.5}{10}$$

$$= \frac{1}{20}$$

Therefore, the variance is $\frac{1}{20}$.

6. Determine the volume of the solid obtained by revolving the curve:

$$f(y) = y^2$$

about the y-axis over the interval $[0, 2]$, using the shell method.

Solution:

$$V = 2\pi \int_0^2 y(y^2)\,dy$$

$$= 2\pi \int_0^2 y^3\,dy$$

$$= 2\pi \left[\frac{y^4}{4} \right]_0^2$$

$$= 2\pi \left[\frac{16}{4} \right]$$

$$= 2\pi \cdot 4$$

$$= 8\pi$$

Therefore, the volume is 8π.

151

Practice Problems 2

1. Compute the area under the function $f(x) = 2x^2 - 3x + 1$ from $x = 0$ to $x = 3$.

2. Determine the volume of the solid obtained by revolving the region under the curve $y = \sqrt{x}$ from $x = 0$ to $x = 4$ about the x-axis using the disk method.

3. Calculate the volume of the solid formed by revolving the area between $y = x^2$ and $y = \sqrt{x}$ from $x = 0$ to $x = 1$ around the x-axis using the washer method.

4. Find the expected value of the continuous random variable X with probability density function $f(x) = 3x^2$ over the interval $[0, 1]$.

5. Evaluate the variance of the continuous random variable X with probability density function $f(x) = 6x(1 - x)$ over interval $[0, 1]$.

6. Compute the volume using the shell method for the solid obtained by rotating $y = \cos(x)$ from $x = 0$ to $x = \pi/2$ about the y-axis.

Answers 2

1. **Solution:**

$$A = \int_0^3 (2x^2 - 3x + 1)\, dx$$

$$= \left[\frac{2}{3}x^3 - \frac{3}{2}x^2 + x \right]_0^3$$

$$= \left(\frac{2}{3}(3)^3 - \frac{3}{2}(3)^2 + 3 \right) - \left(\frac{2}{3}(0)^3 - \frac{3}{2}(0)^2 + 0 \right)$$

$$= (18 - 13.5 + 3)$$

$$= 7.5$$

Therefore, the area is 7.5.

2. **Solution:**

$$V = \pi \int_0^4 (\sqrt{x})^2\, dx = \pi \int_0^4 x\, dx$$

$$= \pi \left[\frac{x^2}{2} \right]_0^4$$

$$= \pi \left(\frac{4^2}{2} - \frac{0^2}{2} \right)$$

$$= \pi \times 8 = 8\pi$$

Hence, the volume is 8π.

3. **Solution:**

$$V = \pi \int_0^1 \left((\sqrt{x})^2 - (x^2)^2 \right)\, dx$$

$$= \pi \int_0^1 (x - x^4)\, dx$$

$$= \pi \left[\frac{x^2}{2} - \frac{x^5}{5} \right]_0^1$$

$$= \pi \left(\frac{1}{2} - \frac{1}{5} \right)$$

$$= \pi \times \frac{3}{10}$$

Thus, the volume is $\frac{3}{10}\pi$.

4. **Solution:**

$$E[X] = \int_0^1 x \cdot 3x^2 \, dx = 3 \int_0^1 x^3 \, dx$$

$$= 3 \left[\frac{x^4}{4} \right]_0^1$$

$$= 3 \times \frac{1}{4}$$

$$= \frac{3}{4}$$

Therefore, the expected value is $\frac{3}{4}$.

5. **Solution:**

$$E[X] = \int_0^1 x \cdot 6x(1-x) \, dx = 6 \left[\int_0^1 (x^2 - x^3) \, dx \right]$$

$$= 6 \left[\frac{x^3}{3} - \frac{x^4}{4} \right]_0^1$$

$$= 6 \left(\frac{1}{3} - \frac{1}{4} \right)$$

$$= 6 \times \frac{1}{12}$$

$$= \frac{1}{2}$$

Now, compute $E[X^2]$:

$$E[X^2] = \int_0^1 x^2 \cdot 6x(1-x) \, dx = 6 \left[\int_0^1 (x^3 - x^4) \, dx \right]$$

$$= 6 \left[\frac{x^4}{4} - \frac{x^5}{5} \right]_0^1$$

$$= 6 \left(\frac{1}{4} - \frac{1}{5} \right)$$

$$= 6 \times \frac{1}{20}$$

$$= \frac{3}{10}$$

Hence, $\text{Var}(X) = E[X^2] - (E[X])^2$

$$\text{Var}(X) = \frac{3}{10} - \left(\frac{1}{2} \right)^2$$

$$= \frac{3}{10} - \frac{1}{4}$$

$$= \frac{3}{10} - \frac{5}{20}$$

$$= \frac{1}{20}$$

The variance is $\frac{1}{20}$.

6. **Solution:**

$$V = 2\pi \int_0^{\pi/2} x \cos(x) \, dx$$

Integration by parts, let $u = x$ and $dv = \cos(x)dx$, then $du = dx$ and $v = \sin(x)$.

$$V = 2\pi \left[x \sin(x)|_0^{\pi/2} - \int_0^{\pi/2} \sin(x)dx \right]$$

$$= 2\pi \left(\frac{\pi}{2} \cdot 1 - [-\cos(x)]_0^{\pi/2} \right)$$

$$= 2\pi \left(\frac{\pi}{2} + 1 \right)$$

$$= \pi^2 + 2\pi$$

Therefore, the volume is $\pi^2 + 2\pi$.

Practice Problems 3

1. Compute the area under the curve of the function $f(x) = 3x^2 + 2x + 1$ from $x = 1$ to $x = 4$.

2. Find the volume of the solid obtained by revolving the region bounded by $y = x^2$ and $y = 0$ around the x-axis from $x = 0$ to $x = 3$.

3. Determine the expected value $E[X]$ of the continuous random variable with probability density function $f(x) = 6x(1-x)$ on the interval $[0, 1]$.

4. Calculate the volume of the solid generated by revolving the region bounded by $y = \sqrt{x}$, $y = 0$, $x = 0$, and $x = 4$ about the y-axis using the shell method.

5. Evaluate the variance $\text{Var}(X)$ of a random variable X with probability density function $f(x) = e^{-x}$ for $x \geq 0$.

6. Using integration, compute the area between the curves $y = x^3$ and $y = x$ over the interval $[0, 1]$.

Answers 3

1. Compute the area under the curve of the function $f(x) = 3x^2 + 2x + 1$ from $x = 1$ to $x = 4$.
   Solution:
   $$A = \int_1^4 (3x^2 + 2x + 1)\, dx$$
   $$= \left[x^3 + x^2 + x \right]_1^4$$
   $$= \left(4^3 + 4^2 + 4 \right) - \left(1^3 + 1^2 + 1 \right)$$
   $$= (64 + 16 + 4) - (1 + 1 + 1)$$
   $$= 84 - 3 = 81$$

   Therefore, the area is 81 square units.

2. Find the volume of the solid obtained by revolving the region bounded by $y = x^2$ and $y = 0$ around the x-axis from $x = 0$ to $x = 3$.
   Solution:
   $$V = \pi \int_0^3 (x^2)^2\, dx = \pi \int_0^3 x^4\, dx$$

$$= \pi \left[\frac{x^5}{5} \right]_0^3$$

$$= \pi \left(\frac{3^5}{5} - \frac{0^5}{5} \right)$$

$$= \pi \cdot \frac{243}{5} = \frac{243\pi}{5}$$

Therefore, the volume is $\frac{243\pi}{5}$ cubic units.

3. Determine the expected value $E[X]$ of the continuous random variable with probability density function $f(x) = 6x(1-x)$ on the interval $[0, 1]$.
   Solution:

$$E[X] = \int_0^1 x \cdot 6x(1-x)\, dx$$

$$= \int_0^1 6x^2 - 6x^3\, dx$$

$$= \left[2x^3 - \frac{3x^4}{2} \right]_0^1$$

$$= \left(2(1)^3 - \frac{3(1)^4}{2} \right) - \left(2(0)^3 - \frac{3(0)^4}{2} \right)$$

$$= \left(2 - \frac{3}{2} \right)$$

$$= \frac{1}{2}$$

Therefore, the expected value is $\frac{1}{2}$.

4. Calculate the volume of the solid generated by revolving the region bounded by $y = \sqrt{x}$, $y = 0$, $x = 0$, and $x = 4$ about the y-axis using the shell method.
   Solution:

$$V = 2\pi \int_0^4 x\sqrt{x}\, dx$$

$$= 2\pi \int_0^4 x^{3/2}\, dx$$

$$= 2\pi \left[\frac{x^{5/2}}{5/2} \right]_0^4$$

$$= 2\pi \cdot \frac{2}{5} \left[x^{5/2} \right]_0^4$$

$$= \frac{4\pi}{5} \left(4^{5/2} - 0 \right)$$

$$= \frac{4\pi}{5} \cdot 32$$

$$= \frac{128\pi}{5}$$

Therefore, the volume is $\frac{128\pi}{5}$ cubic units.

5. Evaluate the variance $\text{Var}(X)$ of a random variable X with probability density function $f(x) = e^{-x}$ for $x \geq 0$.

Solution: First, find $E[X]$:

$$E[X] = \int_0^\infty x e^{-x}\, dx$$

To compute this, use integration by parts with $u = x$ and $dv = e^{-x}\, dx$, hence $du = dx$ and $v = -e^{-x}$.

$$E[X] = -xe^{-x}\Big|_0^\infty + \int_0^\infty e^{-x}\, dx$$

$$= [0 + 0] + \left[-e^{-x}\right]_0^\infty$$

$$= 1$$

Now calculate $E[X^2]$:

$$E[X^2] = \int_0^\infty x^2 e^{-x}\, dx$$

Use integration by parts twice or lookup the Gamma function result:

$$E[X^2] = 2$$

Therefore,

$$\text{Var}(X) = E[X^2] - (E[X])^2 = 2 - 1^2 = 1$$

6. Using integration, compute the area between the curves $y = x^3$ and $y = x$ over the interval $[0, 1]$.

Solution:

$$A = \int_0^1 (x - x^3)\, dx$$

$$= \int_0^1 (x - x^3)\, dx$$

$$= \left[\frac{x^2}{2} - \frac{x^4}{4}\right]_0^1$$

$$= \left(\frac{1^2}{2} - \frac{1^4}{4}\right) - \left(\frac{0^2}{2} - \frac{0^4}{4}\right)$$

$$= \left(\frac{1}{2} - \frac{1}{4}\right)$$

$$= \frac{1}{4}$$

Therefore, the area is $\frac{1}{4}$ square units.

Chapter 16

Multiple Integrals

Practice Problems 1

1. Evaluate the double integral by iterated integration:

$$\iint_{[0,1]\times[0,1]} (x^2 + y^2)\, dy\, dx$$

2. Change the order of integration and evaluate the integral:

$$\int_0^1 \int_{x^2}^1 6xy\, dy\, dx$$

3. Evaluate the double integral over a triangular region:

$$\iint_D (x + y)\, dA$$

where D is bounded by $x = 0$, $y = 0$, and $x + y = 1$.

4. Use a change of variables to evaluate:

$$\iint_D (x^2 + y^2)\, dx\, dy$$

where D is the region bounded by $x^2 + y^2 \leq 1$ using polar coordinates.

5. Evaluate the triple integral:

$$\iiint_{[0,1]\times[0,1]\times[0,1]} xyz\, dz\, dy\, dx$$

6. Compute the probability for the bivariate uniform distribution over a triangular region:

$$P((X,Y) \in D) = \iint_D \frac{1}{2}\, dx\, dy$$

where D is the triangular region with vertices at $(0,0)$, $(1,0)$, and $(0,1)$.

Answers 1

1. Evaluate the double integral by iterated integration:

$$\iint_{[0,1]\times[0,1]} (x^2 + y^2)\, dy\, dx$$

Solution:

$$\int_0^1 \left(\int_0^1 (x^2 + y^2)\, dy \right) dx$$

160

First, integrate with respect to y:

$$\int_0^1 (x^2 + y^2)\, dy = \left[x^2 y + \frac{y^3}{3} \right]_0^1 = x^2 + \frac{1}{3}$$

Next, integrate with respect to x:

$$\int_0^1 \left(x^2 + \frac{1}{3} \right) dx = \left[\frac{x^3}{3} + \frac{x}{3} \right]_0^1 = \frac{1}{3} + \frac{1}{3} = \frac{2}{3}$$

Therefore, the integral evaluates to $\frac{2}{3}$.

2. Change the order of integration and evaluate the integral:

$$\int_0^1 \int_{x^2}^1 6xy\, dy\, dx$$

Solution: The region of integration can be described by $0 \le x \le 1$ and $x^2 \le y \le 1$. In the new order, $0 \le y \le 1$ and $0 \le x \le \sqrt{y}$.

$$\int_0^1 \int_0^{\sqrt{y}} 6xy\, dx\, dy$$

First, integrate with respect to x:

$$\int_0^{\sqrt{y}} 6xy\, dx = \left[3x^2 y \right]_0^{\sqrt{y}} = 3y^2$$

Next, integrate with respect to y:

$$\int_0^1 3y^2\, dy = \left[y^3 \right]_0^1 = 1$$

Therefore, the integral evaluates to 1.

3. Evaluate the double integral over a triangular region:

$$\iint_D (x + y)\, dA$$

where D is bounded by $x = 0$, $y = 0$, and $x + y = 1$.
Solution: The region D is characterized by $0 \le x \le 1$ and $0 \le y \le 1 - x$.

$$\int_0^1 \int_0^{1-x} (x + y)\, dy\, dx$$

Integrate with respect to y:

$$\int_0^{1-x} (x + y)\, dy = \left[xy + \frac{y^2}{2} \right]_0^{1-x} = x(1 - x) + \frac{(1 - x)^2}{2}$$

Simplifying:

$$= x - x^2 + \frac{1 - 2x + x^2}{2} = \frac{1}{2} + x - \frac{3}{2}x^2$$

Integrate with respect to x:

$$\int_0^1 \left(\frac{1}{2} + x - \frac{3}{2}x^2 \right) dx = \left[\frac{x}{2} + \frac{x^2}{2} - \frac{x^3}{2} \right]_0^1$$

$$= \frac{1}{2} + \frac{1}{2} - \frac{1}{2} = \frac{1}{2}$$

Therefore, the integral evaluates to $\frac{1}{2}$.

4. Use a change of variables to evaluate:

$$\iint_D (x^2 + y^2)\, dx\, dy$$

where D is the region bounded by $x^2 + y^2 \leq 1$ using polar coordinates.

Solution: Switching to polar coordinates, $x = r\cos(\theta)$ and $y = r\sin(\theta)$, where $0 \leq r \leq 1$ and $0 \leq \theta \leq 2\pi$.

$$\iint_D (x^2 + y^2)\, dx\, dy = \int_0^{2\pi} \int_0^1 r^2 \cdot r\, dr\, d\theta = \int_0^{2\pi} \int_0^1 r^3\, dr\, d\theta$$

Integrate with respect to r:

$$\int_0^1 r^3\, dr = \left[\frac{r^4}{4} \right]_0^1 = \frac{1}{4}$$

Integrate with respect to θ:

$$\int_0^{2\pi} \frac{1}{4}\, d\theta = \frac{1}{4} \times 2\pi = \frac{\pi}{2}$$

Therefore, the integral evaluates to $\frac{\pi}{2}$.

5. Evaluate the triple integral:

$$\iiint_{[0,1] \times [0,1] \times [0,1]} xyz\, dz\, dy\, dx$$

Solution: Start by integrating with respect to z:

$$\int_0^1 xyz\, dz = \left[\frac{xyz^2}{2} \right]_0^1 = \frac{xy}{2}$$

Then integrate with respect to y:

$$\int_0^1 \frac{xy}{2}\, dy = \left[\frac{xy^2}{4} \right]_0^1 = \frac{x}{4}$$

Finally, integrate with respect to x:

$$\int_0^1 \frac{x}{4}\, dx = \left[\frac{x^2}{8} \right]_0^1 = \frac{1}{8}$$

Therefore, the integral evaluates to $\frac{1}{8}$.

6. Compute the probability for the bivariate uniform distribution over a triangular region:

$$P((X, Y) \in D) = \iint_D \frac{1}{2}\, dx\, dy$$

where D is the triangular region with vertices at $(0,0)$, $(1,0)$, and $(0,1)$.

Solution: The region D is characterized by $0 \leq x \leq 1$ and $0 \leq y \leq 1 - x$.

$$\int_0^1 \int_0^{1-x} \frac{1}{2}\, dy\, dx$$

Integrate with respect to y:

$$\int_0^{1-x} \frac{1}{2}\, dy = \left[\frac{y}{2} \right]_0^{1-x} = \frac{1-x}{2}$$

Integrate with respect to x:

$$\int_0^1 \frac{1-x}{2}\, dx = \left[\frac{x}{2} - \frac{x^2}{4} \right]_0^1 = \frac{1}{2} - \frac{1}{4} = \frac{1}{4}$$

Therefore, the probability evaluates to $\frac{1}{4}$.

Practice Problems 2

1. Evaluate the double integral over the region D defined by $0 \le x \le 2$ and $0 \le y \le x$:

$$\iint_D (x + y) \, dy \, dx$$

2. Compute the triple integral of $f(x, y, z) = xyz$ over the rectangular box $V = [0, 1] \times [0, 1] \times [0, 1]$:

$$\iiint_V xyz \, dx \, dy \, dz$$

3. Using the change of variables, evaluate the double integral:

$$\iint_D (x^2 + y^2) \, dx \, dy$$

where D is the unit disk $x^2 + y^2 \le 1$.

4. Show the transformation steps to evaluate the double integral:

$$\iint_D (2x + 3y) \, dx \, dy$$

where D is bounded by $y = x, y = 0, x = 1$.

5. Determine the probability that a random variable (X, Y) falls within the region D when:

$$f_{XY}(x, y) = 6x(1 - y), \quad 0 \leq x \leq 1, 0 \leq y \leq 1$$

$$\text{where } D = \{(x, y) \mid x + y \leq 1\}$$

6. Calculate the volume under the surface $z = x^2 + y^2$ and above the region R in the xy-plane bounded by $y = x^2$ and $y = 1$.

Answers 2

1. Evaluate the double integral over the region D defined by $0 \leq x \leq 2$ and $0 \leq y \leq x$:

$$\iint_D (x + y) \, dy \, dx$$

Solution: First, we determine the limits for y, which are from 0 to x. Then, the limits for x are from 0 to 2. Therefore, the integral becomes:

$$\int_0^2 \left(\int_0^x (x + y) \, dy \right) dx$$

Evaluating the inner integral with respect to y:

$$\int_0^x (x + y) \, dy = \left[xy + \frac{y^2}{2} \right]_0^x = x^2 + \frac{x^2}{2} = \frac{3}{2} x^2$$

Now, we evaluate the outer integral:

$$\int_0^2 \frac{3}{2} x^2 \, dx = \frac{3}{2} \int_0^2 x^2 \, dx = \frac{3}{2} \left[\frac{x^3}{3} \right]_0^2 = \frac{3}{2} \cdot \frac{8}{3} = 4$$

So, the value of the integral is 4.

2. Compute the triple integral of $f(x, y, z) = xyz$ over the rectangular box $V = [0, 1] \times [0, 1] \times [0, 1]$:

$$\iiint_V xyz \, dx \, dy \, dz$$

Solution: This is a straightforward integration, where each limit ranges from 0 to 1.

$$\int_0^1 \int_0^1 \int_0^1 xyz \, dx \, dy \, dz$$

Start with the integral with respect to x:

$$\int_0^1 x \, dx = \left[\frac{x^2}{2}\right]_0^1 = \frac{1}{2}$$

Now with respect to y:

$$\int_0^1 y \cdot \frac{1}{2} \, dy = \frac{1}{2}\left[\frac{y^2}{2}\right]_0^1 = \frac{1}{2} \cdot \frac{1}{2} = \frac{1}{4}$$

Finally, integrate over z:

$$\int_0^1 z \cdot \frac{1}{4} \, dz = \frac{1}{4}\left[\frac{z^2}{2}\right]_0^1 = \frac{1}{4} \cdot \frac{1}{2} = \frac{1}{8}$$

Therefore, the value of the triple integral is $\frac{1}{8}$.

3. Using the change of variables, evaluate the double integral:

$$\iint_D (x^2 + y^2) \, dx \, dy$$

where D is the unit disk $x^2 + y^2 \leq 1$.

Solution: We use polar coordinates: $x = r\cos\theta$, $y = r\sin\theta$. The Jacobian is r.

$$\iint_D (r^2) \cdot r \, dr \, d\theta = \int_0^{2\pi} \int_0^1 r^3 \, dr \, d\theta$$

Integrate with respect to r:

$$\int_0^1 r^3 \, dr = \left[\frac{r^4}{4}\right]_0^1 = \frac{1}{4}$$

Now with respect to θ:

$$\int_0^{2\pi} \frac{1}{4} \, d\theta = \frac{1}{4} \cdot 2\pi = \frac{\pi}{2}$$

So, the value of the integral is $\frac{\pi}{2}$.

4. Show the transformation steps to evaluate the double integral:

$$\iint_D (2x + 3y) \, dx \, dy$$

where D is bounded by $y = x, y = 0, x = 1$.

Solution: The region D describes a triangular region in the xy-plane with vertices at $(0, 0)$, $(1, 0)$, and $(1, 1)$. The limits for y are from 0 to x, while x ranges from 0 to 1.

$$\int_0^1 \int_0^x (2x + 3y) \, dy \, dx$$

Integrate with respect to y:

$$\int_0^x (2x + 3y) \, dy = \left[2xy + \frac{3y^2}{2}\right]_0^x = 2x^2 + \frac{3x^2}{2} = \frac{7}{2}x^2$$

Now with respect to x:

$$\int_0^1 \frac{7}{2}x^2 \, dx = \frac{7}{2}\left[\frac{x^3}{3}\right]_0^1 = \frac{7}{2} \cdot \frac{1}{3} = \frac{7}{6}$$

Therefore, the integral evaluates to $\frac{7}{6}$.

165

5. Determine the probability that a random variable (X, Y) falls within the region D when:

$$f_{XY}(x, y) = 6x(1 - y), \quad 0 \le x \le 1, 0 \le y \le 1$$

$$\text{where } D = \{(x, y) \mid x + y \le 1\}$$

Solution: The region D is described by the inequality $x + y \le 1$. Solving for y, we have $y \le 1 - x$.

$$\int_0^1 \int_0^{1-x} 6x(1 - y) \, dy \, dx$$

Integrate with respect to y:

$$\int_0^{1-x} 6x(1 - y) \, dy = 6x \left[y - \frac{y^2}{2} \right]_0^{1-x} = 6x \left((1 - x) - \frac{(1 - x)^2}{2} \right)$$

Simplify and evaluate the integral:

$$= 6x \left(1 - x - \frac{1 - 2x + x^2}{2} \right) = 6x \left(\frac{1 - x^2}{2} \right) = 3x(1 - x^2)$$

Integrate this with respect to x:

$$\int_0^1 3x(1 - x^2) \, dx = \int_0^1 (3x - 3x^3) \, dx = \left[\frac{3x^2}{2} - \frac{3x^4}{4} \right]_0^1 = \frac{3}{2} - \frac{3}{4} = \frac{3}{4}$$

Thus, the probability is $\frac{3}{4}$.

6. Calculate the volume under the surface $z = x^2 + y^2$ and above the region R in the xy-plane bounded by $y = x^2$ and $y = 1$.
   Solution: We first think of the limits in y based on the functions: $y = x^2$ to $y = 1$. Along x, this implies the region from 0 to 1 by examining $y = x^2 \to y = 1$ generates an interval in x from 0 to 1.

$$\int_0^1 \int_{x^2}^1 (x^2 + y^2) \, dy \, dx$$

Evaluating the inner integral with respect to y:

$$\int_{x^2}^1 (x^2 + y^2) \, dy = \left[x^2 y + \frac{y^3}{3} \right]_{x^2}^1 = x^2(1 - x^2) + \frac{1 - (x^6)}{3}$$

Simplifying we get:

$$x^2 - x^4 + \frac{1}{3} - \frac{x^6}{3}$$

Integrate this expression with respect to x from 0 to 1:

$$= \int_0^1 (x^2 - x^4 + \frac{1}{3} - \frac{x^6}{3}) \, dx = \left[\frac{x^3}{3} - \frac{x^5}{5} + \frac{x}{3} - \frac{x^7}{21} \right]_0^1$$

Compute after evaluating:

$$= \frac{1}{3} - \frac{1}{5} + \frac{1}{3} - \frac{1}{21} = \frac{7}{21} - \frac{9}{63} + \frac{21}{63} - \frac{3}{63} = \frac{16}{63}$$

Hence, the volume under the surface is $\frac{16}{63}$.

Practice Problems 3

1. Evaluate the double integral of the function $f(x, y) = x^2 + y^2$ over the region defined by $0 \le x \le 1$ and $0 \le y \le 1$.

2. Compute the double integral using polar coordinates for $f(x, y) = x^2 + y^2$ over the unit circle centered at the origin.

3. Change the order of integration and evaluate the integral $\int_0^1 \int_x^1 e^{y^2} \, dy \, dx$.

4. Use a change of variables to evaluate the integral $\iint_D (x + y) \, dx \, dy$, where D is the region bounded by $x = 0$, $y = 0$, and $x + y = 1$.

5. Calculate the triple integral $\iiint_V (2x + y - z) \, dx \, dy \, dz$ over the region V defined by $0 \le x \le 1, 0 \le y \le 1$, and $0 \le z \le 1$.

6. Find the probability that a random variable (X, Y) with joint density function $f_{XY}(x, y) = 6xy$ for $0 \leq x \leq 1$, $0 \leq y \leq 1 - x$ lies in the region $D : x + y \leq 1$.

Answers 3

1. Evaluate the double integral of the function $f(x, y) = x^2 + y^2$ over the region defined by $0 \leq x \leq 1$ and $0 \leq y \leq 1$.

   Solution:
   $$\iint_{0 \leq x \leq 1, 0 \leq y \leq 1} (x^2 + y^2)\, dx\, dy = \int_0^1 \int_0^1 (x^2 + y^2)\, dy\, dx$$

   We first evaluate the inner integral with respect to y:
   $$\int_0^1 (x^2 + y^2)\, dy = \left[x^2 y + \frac{y^3}{3} \right]_0^1 = x^2 \cdot 1 + \frac{1^3}{3} = x^2 + \frac{1}{3}$$

   Next, we evaluate the outer integral:
   $$\int_0^1 \left(x^2 + \frac{1}{3} \right) dx = \int_0^1 x^2\, dx + \int_0^1 \frac{1}{3}\, dx$$
   $$= \left[\frac{x^3}{3} \right]_0^1 + \left[\frac{1}{3} x \right]_0^1 = \frac{1}{3} + \frac{1}{3} = \frac{2}{3}$$

   Therefore, the value of the double integral is $\frac{2}{3}$.

2. Compute the double integral using polar coordinates for $f(x, y) = x^2 + y^2$ over the unit circle centered at the origin.

   Solution: The unit circle has the limits for radius r, $0 \leq r \leq 1$, and angle θ, $0 \leq \theta \leq 2\pi$. In polar coordinates, $x^2 + y^2 = r^2$. The double integral becomes:
   $$\iint_{x^2 + y^2 \leq 1} (x^2 + y^2)\, dx\, dy = \int_0^{2\pi} \int_0^1 r^2 \cdot r\, dr\, d\theta$$
   $$= \int_0^{2\pi} \int_0^1 r^3\, dr\, d\theta$$

   Evaluate the inner integral:
   $$\int_0^1 r^3\, dr = \left[\frac{r^4}{4} \right]_0^1 = \frac{1}{4}$$

   The outer integral then becomes:
   $$\int_0^{2\pi} \frac{1}{4}\, d\theta = \frac{1}{4} [\theta]_0^{2\pi} = \frac{1}{4} \cdot 2\pi = \frac{\pi}{2}$$

   Therefore, the value of the double integral is $\frac{\pi}{2}$.

3. Change the order of integration and evaluate the integral $\int_0^1 \int_x^1 e^{y^2} \, dy \, dx$.

   Solution: The region of integration is bounded by $0 \leq x \leq 1$ and $x \leq y \leq 1$. We change the order to y from 0 to 1 and x from 0 to y.

$$\int_0^1 \int_x^1 e^{y^2} \, dy \, dx = \int_0^1 \int_0^y e^{y^2} \, dx \, dy$$

   Evaluate the inner integral with respect to x:

$$\int_0^y e^{y^2} \, dx = e^{y^2} [x]_0^y = y e^{y^2}$$

   Evaluate the outer integral:

$$\int_0^1 y e^{y^2} \, dy$$

   Use substitution: let $u = y^2$, then $du = 2y \, dy$, $dy = \frac{du}{2y}$:

$$\int_0^1 y e^{y^2} \, dy = \frac{1}{2} \int_0^1 e^u \, du = \frac{1}{2} [e^u]_0^1 = \frac{1}{2}(e - 1)$$

   Therefore, the value of the integral is $\frac{e-1}{2}$.

4. Use a change of variables to evaluate the integral $\iint_D (x + y) \, dx \, dy$, where D is the region bounded by $x = 0$, $y = 0$, and $x + y = 1$.

   Solution: Define the transformation $u = x + y$, $v = x - y$, then

$$x = \frac{u + v}{2}, \quad y = \frac{u - v}{2}$$

   The Jacobian is:

$$J = \begin{vmatrix} \frac{\partial x}{\partial u} & \frac{\partial x}{\partial v} \\ \frac{\partial y}{\partial u} & \frac{\partial y}{\partial v} \end{vmatrix} = \begin{vmatrix} \frac{1}{2} & \frac{1}{2} \\ \frac{1}{2} & -\frac{1}{2} \end{vmatrix} = -\frac{1}{4} - \frac{1}{4} = \frac{1}{2}$$

   The region D becomes $0 \leq u \leq 1$ and $-u \leq v \leq u$.

   The integral becomes:

$$\iint_D (x + y) \, dx \, dy = \int_0^1 \int_{-u}^u u \cdot \frac{1}{2} \, dv \, du$$

   Evaluate the inner integral:

$$\int_{-u}^u \frac{u}{2} \, dv = \frac{u}{2} [v]_{-u}^u = \frac{u}{2}(2u) = u^2$$

   Evaluate the outer integral:

$$\int_0^1 u^2 \, du = \left[\frac{u^3}{3} \right]_0^1 = \frac{1}{3}$$

   Therefore, the value of the integral is $\frac{1}{3}$.

5. Calculate the triple integral $\iiint_V (2x + y - z) \, dx \, dy \, dz$ over the region V defined by $0 \leq x \leq 1, 0 \leq y \leq 1$, and $0 \leq z \leq 1$.

   Solution:

$$\iiint_V (2x + y - z) \, dx \, dy \, dz = \int_0^1 \int_0^1 \int_0^1 (2x + y - z) \, dz \, dy \, dx$$

Evaluate the inner integral with respect to z:

$$\int_0^1 (2x + y - z)\, dz = [2xz + yz - \frac{z^2}{2}]_0^1 = 2x + y - \frac{1}{2}$$

Evaluate the integral with respect to y:

$$\int_0^1 (2x + y - \frac{1}{2})\, dy = [2xy + \frac{y^2}{2} - \frac{y}{2}]_0^1 = 2x + \frac{1}{2} - \frac{1}{2} = 2x$$

Evaluate the outer integral:

$$\int_0^1 2x\, dx = [x^2]_0^1 = 1$$

Therefore, the value of the triple integral is 1.

6. Find the probability that a random variable (X, Y) with joint density function $f_{XY}(x, y) = 6xy$ for $0 \le x \le 1$, $0 \le y \le 1 - x$ lies in the region $D : x + y \le 1$.

Solution: The joint density function $f_{XY}(x, y) = 6xy$ is already defined over the region $D : 0 \le x \le 1$ and $0 \le y \le 1 - x$.

$$P((X, Y) \in D) = \iint_D 6xy\, dy\, dx = \int_0^1 \int_0^{1-x} 6xy\, dy\, dx$$

Evaluate the inner integral:

$$\int_0^{1-x} 6xy\, dy = 6x \left[\frac{y^2}{2}\right]_0^{1-x} = 6x \cdot \frac{(1-x)^2}{2} = 3x(1 - 2x + x^2)$$

Evaluate the outer integral:

$$\int_0^1 (3x - 6x^2 + 3x^3)\, dx = \left[\frac{3x^2}{2} - 2x^3 + \frac{3x^4}{4}\right]_0^1 = \frac{3}{2} - 2 + \frac{3}{4} = \frac{1}{4}$$

Therefore, the probability is $\frac{1}{4}$.

Chapter 17

Differential Equations in Machine Learning

Practice Problems 1

1. Solve the first-order ordinary differential equation (ODE) given by:

$$\frac{dy}{dx} = 2y - 3$$

2. Consider the partial differential equation (PDE) known as the wave equation. Given by:

$$\frac{\partial^2 u}{\partial t^2} = c^2 \frac{\partial^2 u}{\partial x^2}$$

Identify the nature of this PDE and explain its implications in modeling phenomena.

3. Use Euler's method to approximate the solution for the ODE:

$$\frac{dy}{dx} = y + x, \quad y(0) = 1$$

with step size $h = 0.1$ to find $y(0.1)$.

4. Analyze the stability characteristics of Euler's method when applied to the ODE:

$$\frac{dy}{dx} = -ky$$

where $k > 0$.

5. Determine the general solution of the second-order linear homogeneous differential equation:

$$\frac{d^2y}{dx^2} - 4\frac{dy}{dx} + 4y = 0$$

6. Explain how partial differential equations (PDEs) could be utilized to understand weight updates in neural networks, referring specifically to convection-diffusion phenomena.

Answers 1

1. Solve the first-order ordinary differential equation (ODE) given by:

$$\frac{dy}{dx} = 2y - 3$$

Solution: This is a linear first-order ODE. To solve it, we can use separation of variables or an integrating factor. We'll use the integrating factor method here.

Rewrite the equation as:

$$\frac{dy}{dx} - 2y = -3$$

The integrating factor $\mu(x)$ is:

$$\mu(x) = e^{\int -2\,dx} = e^{-2x}$$

Multiply through by the integrating factor:

$$e^{-2x}\frac{dy}{dx} - 2e^{-2x}y = -3e^{-2x}$$

The left-hand side is the derivative of $e^{-2x}y$:

$$\frac{d}{dx}(e^{-2x}y) = -3e^{-2x}$$

Integrate both sides:

$$e^{-2x}y = \int -3e^{-2x}\,dx = \frac{3}{2}e^{-2x} + C$$

Solving for $y(x)$:

$$y(x) = \frac{3}{2} + Ce^{2x}$$

2. Consider the partial differential equation (PDE) known as the wave equation. Given by:

$$\frac{\partial^2 u}{\partial t^2} = c^2 \frac{\partial^2 u}{\partial x^2}$$

Solution: This PDE is classified as a linear, second-order partial differential equation of hyperbolic type. This classification indicates that it models phenomena involving wave propagation, such as vibrations, sound waves, and electromagnetic waves. The parameter c represents the wave speed, and solutions typically involve traveling wave solutions that satisfy initial and boundary conditions.

3. Use Euler's method to approximate the solution for the ODE:

$$\frac{dy}{dx} = y + x, \quad y(0) = 1$$

with step size $h = 0.1$ to find $y(0.1)$.

Solution: Using Euler's method, the iterative formula is:

$$y_{n+1} = y_n + h \cdot f(x_n, y_n)$$

Here, $f(x, y) = y + x$, start with $y_0 = 1$ and $x_0 = 0$.

First iteration to find $y_1 \approx y(0.1)$:

$$y_1 = y_0 + 0.1(y_0 + x_0) = 1 + 0.1(1 + 0) = 1.1$$

4. Analyze the stability characteristics of Euler's method when applied to the ODE:

$$\frac{dy}{dx} = -ky$$

where $k > 0$.

Solution: The exact solution of the equation is $y = Ce^{-kx}$, which is stable, showing exponential decay. Applying Euler's method here:

$$y_{n+1} = y_n + h(-ky_n) = (1 - hk)y_n$$

For stability, $|1 - hk| < 1$, which implies $0 < h < \frac{2}{k}$.

5. Determine the general solution of the second-order linear homogeneous differential equation:

$$\frac{d^2y}{dx^2} - 4\frac{dy}{dx} + 4y = 0$$

Solution: The characteristic equation is $r^2 - 4r + 4 = 0$.

Factoring gives:

$$(r - 2)^2 = 0$$

Thus, $r = 2$ is a double root. The general solution for double roots is:

$$y(x) = (C_1 + C_2 x)e^{2x}$$

6. Explain how partial differential equations (PDEs) could be utilized to understand weight updates in neural networks, referring specifically to convection-diffusion phenomena.

Solution: In the context of neural networks, partial differential equations (PDEs) like the convection-diffusion equation can model the continuous limit of weight changes. The convection term corresponds to the systematic changes driven by gradients (as in backpropagation), while the diffusion term can model noise or stochastic aspects within weight updates. The convection-diffusion framework can potentially provide insights into how different architectural and training choices affect the smoothness and ubiquity of convergence behavior in training neural networks.

Practice Problems 2

1. Solve the following ordinary differential equation:

$$\frac{dy}{dx} = 3y$$

2. Determine the general solution for the first-order linear differential equation:

$$\frac{dy}{dx} + 2y = e^x$$

3. Solve the partial differential equation using separation of variables:

$$\frac{\partial u}{\partial t} = 4\frac{\partial^2 u}{\partial x^2}$$

4. Using Euler's method, approximate the value of $y(0.1)$ for the differential equation:

$$\frac{dy}{dx} = -2y, \quad y(0) = 1$$

with a step size of $h = 0.05$.

5. Show how the Pontryagin's minimum principle can be applied to a simple problem:

$$\dot{x} = u, \quad x(0) = 0, \quad x(1) = 1$$

where u minimizes the cost $\int_0^1 (u^2 + x^2)\, dt$.

6. For a neural ordinary differential equation (NODE), demonstrate a simple example where:

$$\frac{dz}{dt} = -z + \sin(t), \quad z(0) = 1$$

Compute the solution $z(t)$.

Answers 2

1. Solve the following ordinary differential equation:

$$\frac{dy}{dx} = 3y$$

Solution: The equation is a first-order linear differential equation and can be solved by separation of variables:

$$\frac{dy}{y} = 3\,dx$$

Integrating both sides gives:

$$\ln|y| = 3x + C$$

Therefore, exponentiating both sides:

$$y = e^{3x+C} = Ce^{3x}$$

where C is an arbitrary constant.

2. Determine the general solution for the first-order linear differential equation:

$$\frac{dy}{dx} + 2y = e^x$$

Solution: First find the integrating factor, $\mu(x) = e^{\int 2\,dx} = e^{2x}$.

Multiply the entire equation by the integrating factor:

$$e^{2x}\frac{dy}{dx} + 2e^{2x}y = e^{2x}e^x$$

The left side becomes the derivative of a product:

$$\frac{d}{dx}(e^{2x}y) = e^{3x}$$

Integrate both sides:

$$e^{2x}y = \frac{1}{3}e^{3x} + C$$

Solve for y:

$$y = \frac{1}{3}e^x + Ce^{-2x}$$

176

3. Solve the partial differential equation using separation of variables:

$$\frac{\partial u}{\partial t} = 4\frac{\partial^2 u}{\partial x^2}$$

Solution: Assume $u(x,t) = X(x)T(t)$. Substituting into the equation gives:

$$X(x)\frac{dT}{dt} = 4T(t)\frac{d^2 X}{dx^2}$$

Separate variables:

$$\frac{1}{T}\frac{dT}{dt} = 4\frac{1}{X}\frac{d^2 X}{dx^2} = -\lambda$$

For $T(t)$:

$$\frac{dT}{dt} = -\lambda T \quad \Rightarrow \quad T(t) = T_0 e^{-\lambda t}$$

For $X(x)$:

$$\frac{d^2 X}{dx^2} = -\frac{\lambda}{4}X \quad \Rightarrow \quad X(x) = A\cos\left(\frac{\sqrt{\lambda}}{2}x\right) + B\sin\left(\frac{\sqrt{\lambda}}{2}x\right)$$

4. Using Euler's method, approximate the value of $y(0.1)$ for the differential equation:

$$\frac{dy}{dx} = -2y, \quad y(0) = 1$$

Solution: Initial condition: $y_0 = 1$, $h = 0.05$.

For the first step:

$$y_1 = y_0 + h(-2y_0) = 1 + 0.05(-2 \times 1) = 0.9$$

For the second step:

$$y_2 = y_1 + h(-2y_1) = 0.9 + 0.05(-2 \times 0.9) = 0.81$$

Thus, $y(0.1) \approx 0.81$.

5. Show how the Pontryagin's minimum principle can be applied to a simple problem:

$$\dot{x} = u, \quad x(0) = 0, \quad x(1) = 1$$

Solution: The Hamiltonian, H, is defined as $H = u^2 + x^2 + \lambda u$.

Using Pontryagin's principle, minimize H with respect to u:

$$\frac{\partial H}{\partial u} = 2u + \lambda = 0 \quad \Rightarrow \quad u = -\frac{\lambda}{2}$$

The co-state equation:

$$\dot{\lambda} = -\frac{\partial H}{\partial x} = -2x$$

Given $\lambda(1) = 0$, solve the differential equations for x, λ, and u.

6. For a neural ordinary differential equation (NODE), demonstrate a simple example where:

$$\frac{dz}{dt} = -z + \sin(t), \quad z(0) = 1$$

Solution: This is a first-order linear ODE. Solve using integrating factor method:

$$\frac{dz}{dt} + z = \sin(t)$$

with integrating factor $\mu(t) = e^t$:

$$e^t \frac{dz}{dt} + e^t z = e^t \sin(t)$$

Therefore:

$$\frac{d}{dt}(e^t z) = e^t \sin(t)$$

Integrate both sides:

$$e^t z = \int e^t \sin(t)\, dt$$

Use integration by parts: Let $u = \sin(t)$, $dv = e^t dt$, then:

$$\int e^t \sin(t)\, dt = -e^t \cos(t) + e^t \sin(t) + C$$

Solving for $z(t)$:

$$z(t) = -\cos(t) + \sin(t) + Ce^{-t}$$

Initial condition $z(0) = 1$:

$$1 = 0 + 0 + C$$

Hence:

$$z(t) = -\cos(t) + \sin(t) + e^{-t}$$

Practice Problems 3

1. Determine the general solution of the differential equation:

$$\frac{dy}{dx} = 2y$$

2. Solve the following ordinary differential equation using separation of variables:

$$\frac{dy}{dx} = x^2 y$$

3. Verify that the function $y(t) = Ce^{kt}$ is a solution to the following differential equation:

$$\frac{dy}{dt} = ky$$

where C is a constant.

4. Find an analytical solution to the linear differential equation:

$$\frac{dy}{dx} + y = e^x$$

Use the integrating factor method.

5. Consider the partial differential equation from the heat equation:

$$\frac{\partial u}{\partial t} = \alpha \frac{\partial^2 u}{\partial x^2}$$

Discuss whether the separation of variables can be used to solve this PDE.

6. Solve the initial value problem using Euler's method for the differential equation:

$$\frac{dy}{dx} = x + y$$

with initial condition $y(0) = 1$ and step size $h = 0.1$.

Answers 3

1. Determine the general solution of the differential equation:

$$\frac{dy}{dx} = 2y$$

Solution:

$$\frac{dy}{y} = 2\,dx$$

Integrating both sides:

$$\ln|y| = 2x + C$$

$$y = Ce^{2x}$$

Therefore, the general solution is

$$y = Ce^{2x}.$$

2. Solve the following ordinary differential equation using separation of variables:

$$\frac{dy}{dx} = x^2 y$$

Solution:

$$\frac{dy}{y} = x^2\,dx$$

Integrating both sides:

$$\ln|y| = \frac{x^3}{3} + C$$

$$y = Ce^{\frac{x^3}{3}}$$

Therefore, the solution is

$$y = Ce^{\frac{x^3}{3}}.$$

3. Verify that the function $y(t) = Ce^{kt}$ is a solution to the differential equation:

$$\frac{dy}{dt} = ky$$

Solution: Compute the derivative of $y(t)$:

$$\frac{dy}{dt} = \frac{d}{dt}(Ce^{kt}) = Cke^{kt}$$

Substitute $y(t) = Ce^{kt}$:

$$ky = k(Ce^{kt}) = Cke^{kt}$$

Therefore,

$$\frac{dy}{dt} = Cke^{kt} = ky$$

Hence, $y(t) = Ce^{kt}$ satisfies the equation.

4. Find an analytical solution to the linear differential equation:

$$\frac{dy}{dx} + y = e^x$$

Solution: The integrating factor is:

$$\mu(x) = e^{\int 1\,dx} = e^x$$

Multiply through by $\mu(x)$:

$$e^x \frac{dy}{dx} + e^x y = e^{2x}$$

The left side is the derivative of $(e^x y)$:

$$\frac{d}{dx}(e^x y) = e^{2x}$$

Integrate both sides:

$$e^x y = \frac{1}{2} e^{2x} + C$$

$$y = \frac{1}{2} e^x + C e^{-x}$$

Therefore, the solution is

$$y = \frac{1}{2} e^x + C e^{-x}.$$

5. Discuss whether the separation of variables can be used to solve the PDE:

$$\frac{\partial u}{\partial t} = \alpha \frac{\partial^2 u}{\partial x^2}$$

Solution: The separation of variables assumes a solution of the form $u(x,t) = X(x)T(t)$. Substituting:

$$X(x)\frac{dT}{dt} = \alpha \frac{d^2 X}{dx^2} T(t)$$

Dividing both sides by $\alpha X(x)T(t)$:

$$\frac{1}{\alpha T(t)} \frac{dT}{dt} = \frac{1}{X(x)} \frac{d^2 X}{dx^2} = -\lambda$$

This provides two ordinary differential equations:

$$\frac{dT}{dt} + \alpha \lambda T = 0, \quad \frac{d^2 X}{dx^2} + \lambda X = 0$$

These ODEs can be solved individually, verifying that separation of variables is applicable.

6. Solve the initial value problem using Euler's method:

$$\frac{dy}{dx} = x + y, \quad y(0) = 1$$

with step size $h = 0.1$.
Solution:

$$y_{n+1} = y_n + h(x_n + y_n)$$

Initial condition: $y_0 = 1, x_0 = 0$
First step:

$$y_1 = 1 + 0.1(0 + 1) = 1.1$$

$$x_1 = 0 + 0.1 = 0.1$$

Second step:

$$y_2 = 1.1 + 0.1(0.1 + 1.1) = 1.1 + 0.12 = 1.22$$

$$x_2 = 0.1 + 0.1 = 0.2$$

Therefore, after two iterations:

$$y_2 = 1.22$$

Chapter 18

Numerical Methods in Calculus

Practice Problems 1

1. Approximate the derivative of the function $f(x) = \sin(x)$ at $x = \frac{\pi}{4}$ using the central difference method with $h = 0.01$.

2. Calculate the truncation error in the central difference approximation of the derivative for $f(x) = e^x$ at $x = 1$, using $h = 0.01$.

3. Use the trapezoidal rule to approximate the integral of $f(x) = x^2$ over the interval $[0, 1]$ with $n = 2$ subintervals.

4. Apply Simpson's Rule to approximate the integral of $f(x) = \ln(x)$ over the interval $[1, 2]$ with $n = 4$ subintervals.

5. Implement Gaussian quadrature to estimate the integral of $f(x) = x^3$ over $[-1, 1]$ using two-point quadrature.

6. Refine the approximation of the integral of $f(x) = \cos(x)$ over $[0, \frac{\pi}{2}]$ using Romberg integration with initial $R_{1,1}$ from the trapezoidal rule at one subinterval.

Answers 1

1. Approximate the derivative of the function $f(x) = \sin(x)$ at $x = \frac{\pi}{4}$ using the central difference method with $h = 0.01$.

   Solution:
   Using the central difference method:

   $$f'(x) \approx \frac{f(x+h) - f(x-h)}{2h}$$

   $$f'\left(\frac{\pi}{4}\right) \approx \frac{\sin\left(\frac{\pi}{4} + 0.01\right) - \sin\left(\frac{\pi}{4} - 0.01\right)}{2 \times 0.01}$$

   Calculation yields:

   $$\approx \frac{\sin(0.795) - \sin(0.755)}{0.02}$$

   $$\approx \frac{0.713 - 0.678}{0.02} = 1.75$$

The approximation of the derivative is approximately 1.75.

2. Calculate the truncation error in the central difference approximation of the derivative for $f(x) = e^x$ at $x = 1$, using $h = 0.01$.

   Solution:
   The truncation error is proportional to the second derivative $f''(x)$.

   $$f(x) = e^x, \quad f''(x) = e^x$$

   At $x = 1$, $f''(1) = e$. For central difference, the error term is:

   $$\text{Error} \approx -\frac{h^2}{6} f''(x) = -\frac{(0.01)^2}{6} e$$

   $$\approx -\frac{0.0001}{6} \times 2.718 = -0.0000453$$

   The truncation error is approximately -0.0000453.

3. Use the trapezoidal rule to approximate the integral of $f(x) = x^2$ over the interval $[0, 1]$ with $n = 2$ subintervals.

   Solution:
   The interval is divided into 2 subintervals: $[0, 0.5]$ and $[0.5, 1]$.

   $$h = \frac{1 - 0}{2} = 0.5$$

   $$\int_0^1 x^2 \, dx \approx \frac{0.5}{2} [f(0) + 2f(0.5) + f(1)]$$

   $$= 0.25[0 + 2(0.25) + 1]$$

   $$= 0.25[1.5] = 0.375$$

   The approximate integral is 0.375.

4. Apply Simpson's Rule to approximate the integral of $f(x) = \ln(x)$ over the interval $[1, 2]$ with $n = 4$ subintervals.

   Solution:
   Applying Simpson's Rule:

   $$h = (2 - 1)/4 = 0.25$$

   $$\int_1^2 \ln(x) \, dx \approx \frac{h}{3} [f(1) + 4f(1.25) + 2f(1.5) + 4f(1.75) + f(2)]$$

   $$\approx \frac{0.25}{3} [0 + 4(0.2231) + 2(0.4055) + 4(0.5606) + 0.6931]$$

   $$\approx \frac{0.25}{3} [0 + 0.8924 + 0.811 + 2.2424 + 0.6931]$$

   $$\approx \frac{0.25}{3} \times 4.6389 \approx 0.3866$$

   The approximate integral is 0.3866.

5. Implement Gaussian quadrature to estimate the integral of $f(x) = x^3$ over $[-1, 1]$ using two-point quadrature.

Solution:

For two-point Gaussian quadrature:

$$x_1 = -\frac{\sqrt{1/3}}{}, x_2 = \frac{\sqrt{1/3}}{}, \quad w_1 = w_2 = 1$$

$$\int_{-1}^{1} x^3 \, dx \approx \sum_{i=1}^{2} w_i f(x_i)$$

$$= f(-0.57735) + f(0.57735)$$

$$\approx (-0.57735)^3 + (0.57735)^3 = -0.1923 + 0.1923$$

$$\approx 0$$

The integral estimate is 0.

6. Refine the approximation of the integral of $f(x) = \cos(x)$ over $[0, \frac{\pi}{2}]$ using Romberg integration.

Solution:

Starting with trapezoidal rule:

$$R_{1,1} = \frac{\pi/4}{2}[f(0) + f(\pi/2)]$$

$$= \frac{\pi/4}{2}[1 + 0] = \frac{\pi}{8}$$

Next, apply the Romberg formula:

$$R_{2,1} = \frac{\pi/8}{/}2[1 + \sqrt{2}/2]$$

$$\approx \frac{\pi}{12}(1 + 0.7071) \approx \frac{\pi \times 1.7071}{24}$$

$$\approx 0.675$$

Therefore, the refined approximation using Romberg integration is approximately 0.675.

Practice Problems 2

1. Using the forward finite difference method, approximate the derivative of the function $f(x) = x^2 - 4x + 5$ at $x = 2$ with step size $h = 0.1$.

2. Given a function $f(x) = x^3 - 3x + 2$, use the central difference method to approximate $f'(2)$ using $h = 0.01$.

3. Consider $f(x) = \sin(x)$. Approximate the first derivative at $x = \frac{\pi}{4}$ using the backward difference formula with $h = 0.001$.

4. Use the trapezoidal rule to approximate the integral of the function $f(x) = x^2 + 1$ from $x = 0$ to $x = 2$ using 4 equal subintervals.

5. Apply Simpson's rule to estimate the integral of $f(x) = e^x$ from $x = 0$ to $x = 1$ with 4 subintervals.

6. Calculate the integral of $f(x) = \ln(x + 1)$ from $x = 0$ to $x = 2$ using Gaussian quadrature with three nodes.

Answers 2

1. Using the forward finite difference method, approximate the derivative of the function $f(x) = x^2 - 4x + 5$ at $x = 2$ with step size $h = 0.1$.

   Solution:
   $$f'(2) \approx \frac{f(2 + 0.1) - f(2)}{0.1}$$
   $$f(2.1) = (2.1)^2 - 4(2.1) + 5 = 4.41 - 8.4 + 5 = 1.01$$
   $$f(2) = 2^2 - 4 \cdot 2 + 5 = 4 - 8 + 5 = 1$$
   $$f'(2) \approx \frac{1.01 - 1}{0.1} = \frac{0.01}{0.1} = 0.1$$

   Therefore, $f'(2) \approx 0.1$.

2. Given a function $f(x) = x^3 - 3x + 2$, use the central difference method to approximate $f'(2)$ using $h = 0.01$.

   Solution:
   $$f'(2) \approx \frac{f(2 + 0.01) - f(2 - 0.01)}{2 \times 0.01}$$
   $$f(2.01) = (2.01)^3 - 3(2.01) + 2 \approx 8.120601 - 6.03 + 2 = 4.090601$$
   $$f(1.99) = (1.99)^3 - 3(1.99) + 2 \approx 7.880599 - 5.97 + 2 = 3.910599$$
   $$f'(2) \approx \frac{4.090601 - 3.910599}{0.02} = \frac{0.180002}{0.02} = 9.0001$$

   Therefore, $f'(2) \approx 9.0001$.

3. Consider $f(x) = \sin(x)$. Approximate the first derivative at $x = \frac{\pi}{4}$ using the backward difference formula with $h = 0.001$.

   Solution:
   $$f'\left(\frac{\pi}{4}\right) \approx \frac{f\left(\frac{\pi}{4}\right) - f\left(\frac{\pi}{4} - 0.001\right)}{0.001}$$
   $$f\left(\frac{\pi}{4}\right) = \sin\left(\frac{\pi}{4}\right) = \frac{\sqrt{2}}{2}$$
   $$f\left(\frac{\pi}{4} - 0.001\right) = \sin\left(\frac{\pi}{4} - 0.001\right) \approx \sin\left(\frac{\pi}{4}\right) - 0.001 \cos\left(\frac{\pi}{4}\right)$$
   $$\approx \frac{\sqrt{2}}{2} - 0.001 \cdot \frac{\sqrt{2}}{2} = \frac{\sqrt{2}}{2}(1 - 0.001)$$
   $$f'\left(\frac{\pi}{4}\right) \approx \frac{\frac{\sqrt{2}}{2} - \frac{\sqrt{2}}{2}(1 - 0.001)}{0.001} = \frac{0.001 \frac{\sqrt{2}}{2}}{0.001} = \frac{\sqrt{2}}{2}$$

   Therefore, $f'\left(\frac{\pi}{4}\right) \approx \frac{\sqrt{2}}{2}$.

4. Use the trapezoidal rule to approximate the integral of the function $f(x) = x^2 + 1$ from $x = 0$ to $x = 2$ using 4 equal subintervals.

   Solution: Let $n = 4$, $a = 0$, $b = 2$. The step size $h = \frac{b-a}{n} = \frac{2}{4} = 0.5$.

   $$\int_0^2 (x^2 + 1)\, dx \approx \frac{0.5}{2} \left[f(0) + 2f(0.5) + 2f(1) + 2f(1.5) + f(2) \right]$$
   $$f(0) = 0^2 + 1 = 1, \quad f(0.5) = 0.5^2 + 1 = 1.25,$$
   $$f(1) = 1^2 + 1 = 2, \quad f(1.5) = 1.5^2 + 1 = 3.25, \quad f(2) = 2^2 + 1 = 5$$

$$\int_0^2 (x^2 + 1)\, dx \approx \frac{0.5}{2}[1 + 2(1.25) + 2(2) + 2(3.25) + 5]$$

$$= \frac{0.5}{2}[1 + 2.5 + 4 + 6.5 + 5] = 0.25 \times 19 = 4.75$$

Therefore, the approximation is 4.75.

5. Apply Simpson's rule to estimate the integral of $f(x) = e^x$ from $x = 0$ to $x = 1$ with 4 subintervals.

Solution: Let $n = 4$, $a = 0$, $b = 1$. The step size $h = \frac{b-a}{4} = \frac{1}{4} = 0.25$.

$$\int_0^1 e^x\, dx \approx \frac{0.25}{3}[f(0) + 4f(0.25) + 2f(0.5) + 4f(0.75) + f(1)]$$

$$f(0) = e^0 = 1, \quad f(0.25) = e^{0.25} \approx 1.284,$$

$$f(0.5) = e^{0.5} \approx 1.649, \quad f(0.75) = e^{0.75} \approx 2.117, \quad f(1) = e^1 \approx 2.718$$

$$\int_0^1 e^x\, dx \approx \frac{0.25}{3}[1 + 4(1.284) + 2(1.649) + 4(2.117) + 2.718]$$

$$= \frac{0.25}{3}[1 + 5.136 + 3.298 + 8.468 + 2.718] = \frac{0.25}{3}[20.62] = \frac{5.155}{3} \approx 1.718$$

Thus, the approximation is ≈ 1.718.

6. Calculate the integral of $f(x) = \ln(x + 1)$ from $x = 0$ to $x = 2$ using Gaussian quadrature with three nodes.

Solution: For three nodes Gaussian quadrature, the general formula is:

$$\int_a^b f(x)\, dx \approx \sum_{i=1}^n w_i f(x_i)$$

Assume the nodes x_i and weights w_i are determined for the interval $[-1, 1]$, so we adjust for $[0, 2]$.

$$x_i = \frac{b+a}{2} + \frac{b-a}{2}\xi_i, \quad w_i = \frac{b-a}{2}\eta_i$$

Use roots $\xi_1 = -\sqrt{\frac{3}{5}}, \xi_2 = 0, \xi_3 = \sqrt{\frac{3}{5}}$.

Transformed nodes for $[0, 2]$:

$$x_1 = 1 - \sqrt{0.6}, \quad x_2 = 1, \quad x_3 = 1 + \sqrt{0.6}$$

Weights stay same as normalized:

$$w_1 = \frac{5}{9}, \quad w_2 = \frac{8}{9}, \quad w_3 = \frac{5}{9}$$

Calculate:

$$f(x_1) = \ln(1 - \sqrt{0.6} + 1), \quad f(x_2) = \ln(2), \quad f(x_3) = \ln(1 + \sqrt{0.6} + 1)$$

$$\approx \frac{2}{2}\left(\frac{5}{9}f(x_1) + \frac{8}{9}f(x_2) + \frac{5}{9}f(x_3)\right)$$

Evaluate values or use table for quicker, then:

$$\approx \frac{2}{2}\left(\frac{5}{9}(0.582) + \frac{8}{9}(0.693) + \frac{5}{9}(1.098)\right)$$

$$= 0.5555 + 0.616 + 0.61 \approx 1.78$$

Therefore, the approximation is 1.78.

Practice Problems 3

1. Using the central difference formula, numerically approximate the derivative of the function:

$$f(x) = x^2 + 3x + 2$$

at $x = 1$ with step size $h = 0.1$.

2. Consider the function $g(x) = e^x$. Use the trapezoidal rule to approximate the integral of $g(x)$ from $a = 0$ to $b = 1$ with $n = 2$ subintervals.

3. Apply Simpson's rule to estimate the integral:

$$\int_0^2 (4x^2 + x)\, dx$$

with $n = 4$ subintervals.

4. Using Romberg integration, compute the integral:

$$\int_0^1 \sin(x)\, dx$$

up to the second extrapolation level.

5. Determine the central difference error for the approximation of the derivative $f'(x)$ for $f(x) = \cos(x)$ at $x = 0$ with $h = 0.01$.

6. Compute the Gaussian quadrature estimate for the integral:

$$\int_{-1}^{1} x^3 \, dx$$

using $n = 2$ nodes.

Answers 3

1. For the function $f(x) = x^2 + 3x + 2$, apply the central difference formula:

$$f'(1) \approx \frac{f(1 + 0.1) - f(1 - 0.1)}{2 \times 0.1}$$

Calculate $f(1.1) = (1.1)^2 + 3(1.1) + 2 = 1.21 + 3.3 + 2 = 6.51$

$$f(0.9) = (0.9)^2 + 3(0.9) + 2 = 0.81 + 2.7 + 2 = 5.51$$

$$f'(1) \approx \frac{6.51 - 5.51}{0.2} = \frac{1}{0.2} = 5$$

Which matches the analytic derivative $f'(x) = 2x + 3$ at $x = 1$, as analytically $f'(1) = 5$.

2. Using the trapezoidal rule for $g(x) = e^x$ from 0 to 1:

$$\int_0^1 e^x \, dx \approx \frac{1 - 0}{2} \left(e^0 + e^1 \right) = \frac{1}{2}(1 + e) \approx \frac{1}{2}(1 + 2.718) \approx 1.859$$

3. For the integral $\int_0^2 (4x^2 + x) \, dx$ using Simpson's rule: The intervals are $x_0 = 0, x_1 = 0.5, x_2 = 1, x_3 = 1.5, x_4 = 2$, and $h = 0.5$.

$$\int_0^2 (4x^2 + x) \, dx \approx \frac{0.5}{3} \left[(4(0)^2 + 0) + 4(4(0.5)^2 + 0.5) + 2(4(1)^2 + 1) + \right.$$

$$\left. 4(4(1.5)^2 + 1.5) + (4(2)^2 + 2) \right]$$

191

Calculate:

$$= \frac{0.5}{3}[0 + 4(1 + 0.5) + 2(4 + 1) + 4(9 + 1.5) + (16 + 2)]$$

$$= \frac{0.5}{3}[0 + 6 + 10 + 41 + 18] = \frac{0.5}{3} \times 75 = 12.5$$

4. Using Romberg integration for $\int_0^1 \sin(x)\,dx$: First, compute the trapezoidal approximation for $n = 1$:

$$R_{0,0} = \frac{1}{2}(\sin(0) + \sin(1))$$

$$R_{0,0} = \frac{1}{2}(0 + 0.8415) \approx 0.42075$$

For $n = 2$,

$$R_{1,0} = \frac{1}{4}(\sin(0) + 2\sin(0.5) + \sin(1))$$

$$\approx \frac{1}{4}(0 + 2 \times 0.4794 + 0.8415) \approx 0.479425$$

Now apply Richardson extrapolation:

$$R_{0,1} = \frac{4 \times 0.479425 - 0.42075}{3} \approx 0.5403$$

5. Calculate the central difference error for $f(x) = \cos(x)$ at $x = 0$:

$$f'(0) \approx \frac{\cos(0.01) - \cos(-0.01)}{0.02}$$

$$= \frac{0.99995 - 0.99995}{0.02} = 0$$

But the actual derivative $f'(x) = -\sin(x)$ gives $f'(0) = 0$. The central difference method has truncation error of order $O(h^2)$, here essentially zero due to symmetry and small h.

6. Gaussian Quadrature for $\int_{-1}^1 x^3\,dx$ using two-point quadrature (Legendre): The nodes and weights for $n = 2$ are $x_1 = -\frac{\sqrt{1/3}}{,}x_2 = \sqrt{1/3}$ and $w_1 = w_2 = 1$. Evaluating:

$$\int_{-1}^1 x^3\,dx \approx 1 \cdot \left(-\left(\frac{\sqrt{1/3}}{}\right)^3\right) + 1 \cdot \left(\frac{\sqrt{1/3}}{}\right)^3$$

$$= -\frac{1}{3\sqrt{3}} + \frac{1}{3\sqrt{3}} = 0$$

The Gaussian quadrature correctly predicts the zero integral due to symmetry.

Chapter 19

Gradient Descent Method

Practice Problems 1

1. Consider the function $f(\mathbf{x}) = \frac{1}{2}(x_1^2 + x_2^2)$. Derive the gradient descent update rule for this quadratic function.

2. Given a convex function $f(x) = e^x + x^2$, determine the Lipschitz constant L for the gradient $\nabla f(x)$.

3. Assume $f(x) = x^3 - 3x$ is the objective function. Find the local minimum using one iteration of gradient descent starting from $x_0 = 1$, with learning rate $\alpha = 0.1$.

4. Investigate the convergence of gradient descent for the function $f(x) = x^4$. Will the method always converge regardless of the step size α?

5. Using stochastic gradient descent, explain how you would update the parameters for the function $f(\mathbf{x}) = \sum_{i=1}^{5}(x_i - i)^2$ using a mini-batch size of 1.

6. Discuss the effect of using a learning rate $\alpha = 1.5$ for the function $f(x) = x^2$ considering it has a Lipschitz gradient with constant $L = 2$. What would happen to the convergence?

Answers 1

1. **Gradient Descent Update Rule:** To derive the gradient descent update rule for $f(\mathbf{x}) = \frac{1}{2}(x_1^2 + x_2^2)$, we first find the gradient:

$$\nabla f(\mathbf{x}) = \begin{pmatrix} \frac{\partial}{\partial x_1} \frac{1}{2}(x_1^2 + x_2^2) \\ \frac{\partial}{\partial x_2} \frac{1}{2}(x_1^2 + x_2^2) \end{pmatrix} = \begin{pmatrix} x_1 \\ x_2 \end{pmatrix}$$

The update rule is:

$$\mathbf{x}_{k+1} = \mathbf{x}_k - \alpha \nabla f(\mathbf{x}_k) = \mathbf{x}_k - \alpha \begin{pmatrix} x_1 \\ x_2 \end{pmatrix} = \begin{pmatrix} x_1 - \alpha x_1 \\ x_2 - \alpha x_2 \end{pmatrix}$$

2. **Lipschitz Constant:** For the function $f(x) = e^x + x^2$, the gradient is $\nabla f(x) = e^x + 2x$. The Lipschitz constant L is found by considering the second derivative:

$$\nabla^2 f(x) = e^x + 2$$

The maximum value of $e^x + 2$ for $x \in \mathbb{R}$ is unbounded, but for a bounded domain, we consider $L = e^b + 2$ if $x \leq b$.

3. **Local Minimum using Gradient Descent:** For $f(x) = x^3 - 3x$, we compute the gradient:

$$\nabla f(x) = 3x^2 - 3$$

At $x_0 = 1$, using $\alpha = 0.1$:

$$x_1 = x_0 - \alpha \nabla f(x_0) = 1 - 0.1(3(1)^2 - 3) = 1 - 0.1 \cdot 0 = 1$$

Thus, after one iteration, the point remains $x_1 = 1$.

4. **Convergence for $f(x) = x^4$:** The gradient $\nabla f(x) = 4x^3$ suggests potential issues with convergence. Generally, step size needs careful setting as $\alpha > 0$ can lead to divergence due to the higher order polynomial effects slowing descent. Hence, step size must be carefully smaller to assure convergence.

5. **SGD Update Rule for $f(\mathbf{x})$:** With a mini-batch size of 1 for $f(\mathbf{x}) = \sum_{i=1}^{5}(x_i - i)^2$:

$$\text{Gradient for the sample } i : \nabla f_i(x_i) = 2(x_i - i)$$

Update step for a single i:

$$x_i \leftarrow x_i - \alpha \cdot 2(x_i - i)$$

This updates individual parameters considering the mini-batch size.

6. **Effect of Large α on Convergence:** With $f(x) = x^2$ and $L = 2$, using $\alpha = 1.5$:

$$\text{Using } \alpha > \frac{2}{2} = 1,$$

convergence isn't guaranteed. Likely causes oscillations as updates overshoot the minimum.

Thus, the learning rate is too large, causing divergence.

Practice Problems 2

1. Show that the gradient descent update rule decreases the function value by proving that:

$$f(\mathbf{x}_{k+1}) \leq f(\mathbf{x}_k) - \frac{\alpha}{2}\|\nabla f(\mathbf{x}_k)\|^2$$

This assumes that f has a Lipschitz continuous gradient with constant L.

2. Explain the impact of the learning rate α on the convergence rate of gradient descent and derive the condition for convergence in terms of α and the Lipschitz constant L.

3. Derive the stochastic gradient descent update rule for a dataset $\mathcal{D}$ and explain how it differs from the batch gradient descent update rule.

4. Prove that for a convex function f, if $\nabla f(\mathbf{x}) = 0$, then $\mathbf{x}$ is a global minimum.

5. Given an objective function $f(\mathbf{x}) = x_1^2 + 2x_2^2$, compute a single step of gradient descent with initial point $\mathbf{x}_0 = (1,1)$ and a learning rate $\alpha = 0.1$.

6. In the context of gradient descent, explain the role of the Hessian matrix and how it can be used to modify the update rules, specifically referencing Newton's method.

Answers 2

1. To show that the gradient descent update rule decreases the function value, we start with the Taylor expansion:

$$f(\mathbf{x}_{k+1}) = f(\mathbf{x}_k - \alpha \nabla f(\mathbf{x}_k)) \approx f(\mathbf{x}_k) - \alpha \nabla f(\mathbf{x}_k)^\top \nabla f(\mathbf{x}_k) + \frac{L}{2} \alpha^2 \|\nabla f(\mathbf{x}_k)\|^2$$

   Since f has a Lipschitz continuous gradient, we have:

$$f(\mathbf{x}_{k+1}) \leq f(\mathbf{x}_k) - \frac{\alpha}{2} \|\nabla f(\mathbf{x}_k)\|^2$$

   This shows that $f(\mathbf{x}_{k+1})$ is less than or equal to $f(\mathbf{x}_k) - \frac{\alpha}{2} \|\nabla f(\mathbf{x}_k)\|^2$, proving the decrease condition.

2. The learning rate α directly influences the convergence rate. For convergence, the choice of α needs to satisfy:

$$0 < \alpha < \frac{2}{L}$$

   Here, L is the Lipschitz constant of the gradient of f. If α is too large, it may cause divergence.

3. For stochastic gradient descent (SGD), the update rule is:

$$\mathbf{x}_{k+1} = \mathbf{x}_k - \alpha \nabla f_{\mathcal{B}}(\mathbf{x}_k)$$

   In SGD, the gradient $\nabla f_{\mathcal{B}}$ is estimated over a mini-batch $\mathcal{B} \subset \mathcal{D}$, which contrasts with batch gradient descent that uses the entire dataset $\mathcal{D}$.

4. In convex optimization, if $\nabla f(\mathbf{x}) = 0$, then by the first-order condition for convex functions, $\mathbf{x}$ is a global minimum. This follows from:

$$f(\mathbf{y}) \geq f(\mathbf{x}) + \nabla f(\mathbf{x})^\top (\mathbf{y} - \mathbf{x})$$

   Since $\nabla f(\mathbf{x}) = 0$, for all $\mathbf{y}$, $f(\mathbf{y}) \geq f(\mathbf{x})$.

5. Given $f(\mathbf{x}) = x_1^2 + 2x_2^2$, the gradient is:

$$\nabla f(\mathbf{x}) = (2x_1, 4x_2)$$

   At $\mathbf{x}_0 = (1, 1)$, it is $(2, 4)$. Taking a step gives:

$$\mathbf{x}_{k+1} = \mathbf{x}_0 - 0.1 \times (2, 4) = (1, 1) - (0.2, 0.4) = (0.8, 0.6)$$

6. The Hessian matrix H provides second-order derivative information. In Newton's method, the update rule is:

$$\mathbf{x}_{k+1} = \mathbf{x}_k - H^{-1} \nabla f(\mathbf{x}_k)$$

   Using the Hessian allows curvature adjustment, potentially accelerating convergence by modifying how step sizes are calculated based on curvature information.

Practice Problems 3

1. Consider a differentiable function $f : \mathbb{R}^n \to \mathbb{R}$ with a Lipschitz continuous gradient. Demonstrate that the function is convex given that there exists a constant $L > 0$ such that for all $\mathbf{x}, \mathbf{y} \in \mathbb{R}^n$,

$$\|\nabla f(\mathbf{x}) - \nabla f(\mathbf{y})\| \leq L \|\mathbf{x} - \mathbf{y}\|$$

2. Derive the convergence condition for the gradient descent method applied to a convex, differentiable function using the update rule:

$$\mathbf{x}_{k+1} = \mathbf{x}_k - \alpha \nabla f(\mathbf{x}_k)$$

with the learning rate α.

3. For a function $f(\mathbf{x}) = \frac{1}{2}\mathbf{x}^\top \mathbf{A}\mathbf{x} + \mathbf{b}^\top \mathbf{x} + c$, where $\mathbf{A}$ is a symmetric positive definite matrix, compute the gradient and establish its role in the gradient descent update rule.

4. Show that for a quadratic function $f(\mathbf{x})$, the iteration of gradient descent converges linearly when the learning rate is chosen based on the largest eigenvalue of the Hessian matrix.

5. Analyze why the choice of learning rate α directly affects the stability and speed of convergence for gradient descent methods. Provide a mathematical explanation of the consequences of an unsuitable learning rate.

6. Explain how Stochastic Gradient Descent (SGD) differs from traditional Gradient Descent in the context of large-scale machine learning problems and the implications for convergence properties.

Answers 3

1. **Convexity via Lipschitz Continuity:**
   Assume f has a Lipschitz continuous gradient with constant L. For convexity:

   $$f(\mathbf{y}) \geq f(\mathbf{x}) + \nabla f(\mathbf{x})^\top (\mathbf{y} - \mathbf{x})$$

   Start from the definition of Lipschitz continuity:

   $$\|\nabla f(\mathbf{x}) - \nabla f(\mathbf{y})\| \leq L \|\mathbf{x} - \mathbf{y}\|$$

   Integrate the gradient from $\mathbf{x}$ to $\mathbf{y}$ along the line segment joining them:

   $$f(\mathbf{y}) - f(\mathbf{x}) \leq \nabla f(\mathbf{x})^\top (\mathbf{y} - \mathbf{x}) + \frac{L}{2} \|\mathbf{y} - \mathbf{x}\|^2$$

   With Dividing, prove convexity.

2. **Convergence Condition for Gradient Descent:**
   To ensure convergence,

   $$f(\mathbf{x}_{k+1}) \leq f(\mathbf{x}_k) - \frac{\alpha}{2} \|\nabla f(\mathbf{x}_k)\|^2$$

   requires optimal α:

   $$0 < \alpha < \frac{2}{L}$$

   due to quadratic upper bound provided by Lipschitz condition.

3. **Gradient for Quadratic Function:**
   Given:

   $$f(\mathbf{x}) = \frac{1}{2} \mathbf{x}^\top \mathbf{A} \mathbf{x} + \mathbf{b}^\top \mathbf{x} + c$$

   the gradient:

   $$\nabla f(\mathbf{x}) = \mathbf{A}\mathbf{x} + \mathbf{b}$$

   Apply in gradient descent $\mathbf{x}_{k+1} = \mathbf{x}_k - \alpha(\mathbf{A}\mathbf{x}_k + \mathbf{b})$.

4. **Linear Convergence for Quadratic Functions:**
   Hessian is $\mathbf{A}$, positive definite. For linear convergence, use eigenvalue:

   $$\alpha = \frac{2}{\lambda_{\max}}$$

   Linear convergence when $\mathbf{A}$ bounds rate: focus on eigenvalue distribution.

5. **Impact of Learning Rate:**
   With α too small, iterative steps take long to convergence. Too large, causes oscillation:

   $$\alpha > \frac{2}{L} \quad \text{not preferred}$$

   Analytically show impact on error sequence relation.

6. **Stochastic Gradient Descent Explanation:**
   SGD updates using:

   $$\nabla f_{\mathcal{B}}(\mathbf{x}_k) = \frac{1}{|\mathcal{B}|} \sum_{i \in \mathcal{B}} \nabla f_i(\mathbf{x}_k)$$

   Benefits and convergence properties for large data: noise reduction and speed.

Chapter 20

Stochastic Gradient Descent

Practice Problems 1

1. Discuss how the expectation and variance properties of stochastic gradients affect the convergence behavior of the stochastic gradient descent (SGD) algorithm.

2. Given a differentiable function $f : \mathbb{R}^n \rightarrow \mathbb{R}$, illustrate how the choice of a mini-batch size $\mathcal{B}$ may influence the performance of the SGD algorithm.

3. Explain why a decaying learning rate is often used in SGD and derive the condition for the convergence of the learning rate schedule $\alpha_k = \frac{\alpha_0}{1+k}$.

4. Consider an initial point $\mathbf{x}_0$, describe how stochastic gradient descent can escape shallow local minima compared to full batch gradient descent. Give a mathematical justification based on variance in updates.

5. Given the SGD update rule $\mathbf{x}_{k+1} = \mathbf{x}_k - \alpha \nabla f_{\mathcal{B}}(\mathbf{x}_k)$, analyze its performance under non-convex objectives and elaborate on any challenges that may arise.

6. For an optimization task with $f(\mathbf{x}) = \sum_{i=1}^{n}(x_i^2 - 4x_i + 4)$, derive the SGD update rule and discuss how the choice of learning rate affects convergence with an illustrative example.

Answers 1

1. **Expectation and Variance in SGD:**

$$\mathbb{E}[\nabla f_{\mathcal{B}}(\mathbf{x})] = \nabla f(\mathbf{x})$$

Solution:
The expectation property indicates that the mini-batch gradient is an unbiased estimator of the true gradient. Variance in stochastic gradients introduces noise in updates that allows the exploration of the parameter space. This exploration helps in escaping shallow local minima, aiding convergence in complex landscapes. However, high variance might also lead to divergence if not well-regulated by the learning rate.

2. **Mini-batch Size Influence:**
   Solution:

The mini-batch size $\mathcal{B}$ affects both computation time and gradient variance. Smaller mini-batches result in higher variance, introducing noise that can traverse flatter regions quickly but may lead to convergence instability. Larger mini-batches reduce variance and stabilize convergence at the expense of computation. Thus, $\mathcal{B}$ balances signal and noise in learning rates.

3. **Decaying Learning Rate:**

$$\alpha_k = \frac{\alpha_0}{1 + k}$$

Solution:
A decaying learning rate reduces step size over time, enabling SGD to fine-tune solutions after exploratory phases. The formal condition for convergence in SGD requires the sum $\sum \alpha_k^2 < \infty$ and the sum of learning rates $\sum \alpha_k = \infty$, ensuring infinite descent under diminishing steps to stabilize near minima.

4. **Escaping Shallow Local Minima:**
Solution:
SGD's stochastic nature means that batches may suggest varied descent directions due to noise. This variation can push the iterate out of shallow local minima, a benefit not offered by deterministic gradients like full-batch gradient descent, which may remain trapped due to consistent update directions.

5. **Non-convex Objective Performance:**
Solution:
In non-convex objectives, SGD's randomness might both aid in escaping local traps and cause oscillations in the search space. This results in a challenge where convergence to a global minimum isn't guaranteed, demanding careful hyperparameter tuning and potential adjustments through adaptive learning techniques.

6. **Optimization with given function:**

$$f(\mathbf{x}) = \sum_{i=1}^{n} (x_i^2 - 4x_i + 4)$$

Solution:
The gradient is $\nabla f(\mathbf{x}) = [2x_i - 4]_{i=1}^{n}$. The SGD update becomes:

$$x_{k+1,i} = x_{k,i} - \alpha(2x_{k,i} - 4)$$

Larger learning rates position the algorithm aggressively altering variables, risking oscillation, while smaller rates ensure stability at the cost of speed. An optimal α achieves a balance, e.g., $\alpha = 0.1$ for steady convergence upon testing for specific empirically derived datasets.

Practice Problems 2

1. Given the differentiable function $f(\mathbf{x}) = \frac{1}{2}\mathbf{x}^T \mathbf{A} \mathbf{x} - \mathbf{b}^T \mathbf{x} + c$, where $\mathbf{x} \in \mathbb{R}^n$, $\mathbf{A}$ is a symmetric positive definite matrix, $\mathbf{b}$ is a vector, and c is a constant, find the gradient $\nabla f(\mathbf{x})$.

2. Consider a function $f : \mathbb{R}^n \to \mathbb{R}$ where $f(\mathbf{x}) = \sum_{i=1}^{n} x_i^2$. What is the Hessian matrix $\mathbf{H}$ of this function?

3. For a logistic regression model, the cost function is given by
$J(\mathbf{w}) = -\frac{1}{m} \sum_{i=1}^{m} \left[y^{(i)} \log(h_{\mathbf{w}}(x^{(i)})) + (1 - y^{(i)}) \log(1 - h_{\mathbf{w}}(x^{(i)})) \right]$, where $h_{\mathbf{w}}(\mathbf{x}) = \frac{1}{1+\exp(-\mathbf{w}^T\mathbf{x})}$. Compute the gradient of $J(\mathbf{w})$ with respect to $\mathbf{w}$.

4. Prove that stochastic gradient descent is an unbiased estimator of the full gradient descent step.

5. Show that if the learning rate α is too large, the iterative SGD update $\mathbf{x}_{k+1} = \mathbf{x}_k - \alpha \nabla f_{\mathcal{B}}(\mathbf{x}_k)$ can diverge.

6. Assuming a linear regression model $y = \mathbf{w}^T\mathbf{x} + b$, derive the closed form solution for the weight vector $\mathbf{w}$ using the normal equation derived from the least squares cost function.

Answers 2

1. Given the differentiable function $f(\mathbf{x}) = \frac{1}{2}\mathbf{x}^T\mathbf{A}\mathbf{x} - \mathbf{b}^T\mathbf{x} + c$.

   Solution:
   $$\nabla f(\mathbf{x}) = \nabla\left(\frac{1}{2}\mathbf{x}^T\mathbf{A}\mathbf{x}\right) - \nabla(\mathbf{b}^T\mathbf{x}) + \nabla c$$

   Using the property of derivatives of quadratic forms:
   $$\nabla\left(\frac{1}{2}\mathbf{x}^T\mathbf{A}\mathbf{x}\right) = \mathbf{A}\mathbf{x}$$

   and
   $$\nabla(\mathbf{b}^T\mathbf{x}) = \mathbf{b}$$
   $$\nabla c = 0$$

   Therefore,
   $$\nabla f(\mathbf{x}) = \mathbf{A}\mathbf{x} - \mathbf{b}$$

2. For the function $f(\mathbf{x}) = \sum_{i=1}^{n} x_i^2$.

   Solution: The second derivative of each term x_i^2 with respect to variables x_i in vector $\mathbf{x}$ will contribute to the diagonal elements of the Hessian.
   $$\mathbf{H} = \begin{bmatrix} 2 & 0 & \cdots & 0 \\ 0 & 2 & \cdots & 0 \\ \vdots & \vdots & \ddots & \vdots \\ 0 & 0 & \cdots & 2 \end{bmatrix} = 2\mathbf{I}$$

   where $\mathbf{I}$ is the identity matrix of size n.

3. For the logistic regression cost function.

   Solution: Start by differentiating the cost function $J(\mathbf{w})$.
   $$\nabla J(\mathbf{w}) = -\frac{1}{m}\sum_{i=1}^{m}\left[\frac{y^{(i)}}{h_{\mathbf{w}}(x^{(i)})}h_{\mathbf{w}}(x^{(i)})(1 - h_{\mathbf{w}}(x^{(i)}))(-x^{(i)}) + \frac{1 - y^{(i)}}{1 - h_{\mathbf{w}}(x^{(i)})}(-h_{\mathbf{w}}(x^{(i)}))(1 - h_{\mathbf{w}}(x^{(i)}))x^{(i)}\right]$$

   Simplify to:
   $$= -\frac{1}{m}\sum_{i=1}^{m}\left[(y^{(i)} - h_{\mathbf{w}}(x^{(i)}))x^{(i)}\right]$$

4. Stochastic gradient descent is an unbiased estimator.

   Solution: We show the expectation of the stochastic gradient equals the full gradient:
   $$\mathbb{E}[\nabla f_{\mathcal{B}}(\mathbf{x})] = \mathbb{E}\left[\frac{1}{|\mathcal{B}|}\sum_{i\in\mathcal{B}}\nabla f_i(\mathbf{x})\right] = \frac{1}{|\mathcal{B}|}\cdot|\mathcal{B}|\cdot\nabla f(\mathbf{x}) = \nabla f(\mathbf{x})$$

   where $f_i(\mathbf{x})$ represents the contribution of each data point.

5. Show divergence if α is too large.

   Solution: Consider a simple case:
   $$\mathbf{x}_{k+1} = \mathbf{x}_k - \alpha\nabla f(\mathbf{x}_k)$$

   If α large, consider a quadratic $f(x) = ax^2 + bx + c$, derivative $\nabla f(x) = 2ax + b$. An overly large α:
   $$\mathbf{x}_{k+1} = \mathbf{x}_k - \alpha(2ax + b)$$

   causes overshooting the minimum if α does not respect Lipschitz conditions or line search considerations, leading to large oscillations or exponential growth in errors.

6. Closed form solution for linear regression.

   Solution: Start with the cost function for least squares:

   $$J(\mathbf{w}) = \|\mathbf{y} - \mathbf{X}\mathbf{w}\|^2$$

   To minimize, set the gradient to zero:

   $$\nabla J(\mathbf{w}) = -2\mathbf{X}^T(\mathbf{y} - \mathbf{X}\mathbf{w}) = 0$$

   Simplify to form the normal equation:

   $$\mathbf{X}^T\mathbf{X}\mathbf{w} = \mathbf{X}^T\mathbf{y}$$

   Solving yields:

   $$\mathbf{w} = (\mathbf{X}^T\mathbf{X})^{-1}\mathbf{X}^T\mathbf{y}$$

Practice Problems 3

1. Consider the function $f : \mathbb{R}^n \to \mathbb{R}$ where $\nabla f(\mathbf{x})$ is available. Discuss how Stochastic Gradient Descent (SGD) modifies the gradient descent update rule to improve computational efficiency with large datasets.

2. Prove the unbiasedness of the stochastic gradient used in SGD by showing that the expectation of the stochastic gradient can be equated to the full gradient.

3. Demonstrate how the choice of a decaying learning rate α_k in SGD ensures convergence, particularly in comparison to a constant learning rate.

4. Using pseudo-code, describe how an SGD algorithm samples mini-batches for gradient estimation and performs parameter updates.

5. Analyze the impact of mini-batch size on the variance of gradient estimates in SGD and how it influences parameter updates.

6. Explain the role of adaptive learning rate strategies in SGD and how they can mitigate the issues associated with poor learning rate selection.

Answers 3

1. **Solution:**

   Stochastic Gradient Descent (SGD) modifies the classic gradient descent algorithm by updating parameters using a randomly selected subset of the dataset, called a mini-batch. This approach mitigates the computational load caused by using the full dataset to compute the gradient. The SGD update step is:

   $$\mathbf{x}_{k+1} = \mathbf{x}_k - \alpha \nabla f_{\mathcal{B}}(\mathbf{x}_k)$$

   By doing so, SGD reduces the time required per iteration, allowing optimal solutions to be reached more efficiently compared to full-batch gradient descent, especially in large datasets.

2. **Solution:**

   The stochastic gradient $\nabla f_{\mathcal{B}}(\mathbf{x})$ is an unbiased estimator of the full gradient $\nabla f(\mathbf{x})$. This unbiasedness can be demonstrated as:

$$\mathbb{E}[\nabla f_{\mathcal{B}}(\mathbf{x})] = \nabla f(\mathbf{x})$$

Here, $\mathbb{E}$ denotes expectation over all possible mini-batches $\mathcal{B}$. Since each mini-batch contributes to the gradient estimation, the expected value across all mini-batches equals the full gradient, maintaining unbiasedness.

3. **Solution:**

A decaying learning rate α_k ensures SGD's convergence by gradually reducing the step size in parameter updates as follows:

$$\alpha_k = \frac{\alpha_0}{1+k}$$

where α_0 is the initial learning rate. This decay scheme prevents overshooting minima in later stages of convergence. In contrast, a constant learning rate might cause erratic updates, leading to oscillations around minima or divergence, particularly in narrow or sharp regions of the loss surface.

4. **Solution:**

The pseudo-code for the SGD algorithm is:

Algorithm 1: Stochastic Gradient Descent

Input: Initial point $\mathbf{x}_0$, learning rate schedule $\{\alpha_k\}$, number of iterations K
for $k = 0, 1, \ldots, K-1$ **do**
    Sample a mini-batch $\mathcal{B}$ from dataset;
    Compute $\mathbf{g}_k = \nabla f_{\mathcal{B}}(\mathbf{x}_k)$;
    Update $\mathbf{x}_{k+1} = \mathbf{x}_k - \alpha_k \mathbf{g}_k$;
return $\mathbf{x}_K$

Mini-batches are randomly selected subsets of the dataset, allowing consistent gradient estimates while adding variability that aids exploration in the optimization landscape.

5. **Solution:**

The mini-batch size affects the variance of gradient estimates. Smaller mini-batches lead to higher variance and noisier updates, which aids in escaping local minima but may hinder convergence. Conversely, larger mini-batches provide stable updates with lower variance but increase computational cost per iteration. Balancing batch size is crucial for efficient SGD performance.

6. **Solution:**

Adaptive learning rate strategies in SGD, such as AdaGrad, RMSProp, or Adam, dynamically adjust the learning rate based on the past gradients. This adjustment helps in tuning the learning rate automatically, mitigating issues of a fixed rate being too large (causing divergence) or too small (leading to slow convergence). Adaptive methods allow SGD to perform well across varying scales and features of the optimization landscape.

Chapter 21

Regularization Techniques

Practice Problems 1

1. Explain the effect of the regularization parameter λ on the L2 regularization term of the cost function and how it impacts model training.

2. Consider a linear regression model using L1 regularization. Why might some coefficients be driven to zero, and what is the geometric interpretation of this phenomenon?

3. For a given linear model utilizing L2 regularization, derive the modified gradient descent update rule.

4. Discuss how L1 regularization can be interpreted as a convex optimization problem, particularly in terms of the effect of non-differentiability at zero.

5. Compare and contrast the stability and sparsity properties of models trained with L1 and L2 regularization.

6. Derive the gradient of the L2 regularized cost function for a model with weights $\mathbf{w}$ and cost function $J(\mathbf{w})$.

Answers 1

1. **Effect of the regularization parameter λ:** The regularization parameter λ in the L2 regularization term $\frac{\lambda}{2}\sum_{j=1}^{n} w_j^2$ controls how much regularization is applied to the model. A larger λ increases the penalty on large weights, therefore discouraging complexity in the model and promoting smaller weight values. This can help reduce overfitting by ensuring the model does not capture noise in the data. Conversely, a smaller λ allows more flexibility but can risk overfitting.

2. **L1 regularization and zero coefficients:** L1 regularization adds a penalty term $\lambda\sum_{j=1}^{n} |w_j|$ to the cost function, which can drive some coefficients exactly to zero. This is because L1 regularization effectively performs feature selection by preferring solutions at the vertex of constraints in the parameter space (which are axis-aligned), leading to sparsity. Geometrically, the L1 regularization has diamond-shaped contours that intersect with the axes, promoting sparse solutions.

3. **Modified gradient descent for L2 regularization:** The standard gradient descent update rule is $\mathbf{w}_j := \mathbf{w}_j - \alpha\frac{\partial J}{\partial \mathbf{w}_j}$. With L2 regularization, the cost function becomes $J_{\text{L2}}(\mathbf{w}) = J(\mathbf{w}) + \frac{\lambda}{2}\sum_{j=1}^{n} w_j^2$. The gradient of the regularization term is λw_j. Therefore, the update rule becomes:

$$\mathbf{w}_j := \mathbf{w}_j - \alpha\left(\frac{\partial J}{\partial \mathbf{w}_j} + \lambda\mathbf{w}_j\right)$$

4. **L1 regularization as a convex optimization problem:** L1 regularization introduces non-differentiability at zero due to the absolute value function $|w_j|$. This makes the optimization problem challenging, but still convex, because the overall objective function remains convex. In optimization, subgradients are often used to handle this non-smoothness. The convex nature ensures that even though $|w_j|$ is not differentiable at zero, a global optimum can still be achieved through numerical optimization techniques.

5. **Stability and sparsity of L1 vs. L2:** Models trained with L1 regularization are more likely to be sparse; many of the coefficients may be zero, effectively choosing only a subset of variables that are most influential. On the other hand, L2 regularization tends to lead to solutions where coefficients are

more evenly distributed among features, providing stability but less sparsity. L2 regularization may be preferred when all features potentially contribute to the model's prediction, whereas L1 is useful for feature selection.

6. **Gradient of L2 regularized cost function:** The cost function with L2 regularization is given by:

$$J_{L2}(\mathbf{w}) = J(\mathbf{w}) + \frac{\lambda}{2} \sum_{j=1}^{n} w_j^2$$

Taking the gradient with respect to w_j:

$$\nabla J_{L2}(\mathbf{w}) = \nabla J(\mathbf{w}) + \lambda \mathbf{w}$$

The regularization term $\frac{\lambda}{2} \sum_{j=1}^{n} w_j^2$ contributes λw_j to each gradient component, thus modifying the slope of the cost function to include the penalty for large weights, promoting smoother and simpler models.

Practice Problems 2

1. Explain the difference between L1 and L2 regularization with respect to their effect on model coefficients.

2. Given the function $J(\mathbf{w}) = \frac{1}{2m} \sum_{i=1}^{m} (y^{(i)} - \mathbf{w}^T \mathbf{x}^{(i)})^2$, derive the gradient descent update rule incorporating L2 regularization.

3. Why is L1 regularization often associated with sparsity in model coefficients? Provide a mathematical interpretation.

4. Describe the optimization landscape when L1 regularization is applied to a convex cost function. How does it differ from L2 regularization?

5. If the learning rate α in a gradient descent algorithm is too large, what effect does it have when L2 regularization is applied? Explain using calculus.

6. Consider the regularization term $\|\mathbf{x}\|_1$. Explain why this term is non-differentiable at zero and how subgradient methods can be employed to optimize L1-regularized problems.

Answers 2

1. The primary difference between L1 and L2 regularization is in their influence on the coefficients of the model. L1 regularization adds a penalty equal to the absolute value of the magnitude of coefficients, $\lambda \sum |w_j|$, and encourages sparsity, often setting some coefficients to zero, effectively performing variable selection. On the other hand, L2 regularization adds a penalty equal to the square of the magnitude of coefficients, $\lambda \sum w_j^2$, discouraging large weights but not eliminating them completely, hence promoting model stability rather than sparsity.

2. Given the function:

$$J(\mathbf{w}) = \frac{1}{2m} \sum_{i=1}^{m} (y^{(i)} - \mathbf{w}^T \mathbf{x}^{(i)})^2$$

The update rule with L2 regularization involves the cost function:

$$J_{\text{L2}}(\mathbf{w}) = J(\mathbf{w}) + \frac{\lambda}{2} \sum_{j=1}^{n} w_j^2$$

The gradient of the regularized cost function is:

$$\nabla J_{\mathrm{L2}}(\mathbf{w}) = \nabla J(\mathbf{w}) + \lambda \mathbf{w}$$

Therefore, the gradient descent update rule becomes:

$$\mathtt{w}_j := \mathtt{w}_j - \alpha \left(\frac{\partial J}{\partial \mathtt{w}_j} + \lambda \mathtt{w}_j \right)$$

The regularization term modifies the update by reducing the weights in each step proportionally to their size and the strength of λ.

3. L1 regularization is associated with sparsity because its penalty term $\lambda \sum |w_j|$ can shrink some coefficients exactly to zero. Mathematically, this is because the absolute value function has a constant gradient except at zero, where it is non-differentiable. This non-differentiable nature at zero allows for the possibility of coefficients being reduced completely to zero during optimization.

4. In the optimization landscape, L1 regularization introduces non-smoothness due to its absolute value penalty, resulting in a geometry that resembles a diamond shape around the origin. This can lead to sparse solutions as optimizer paths are 'pulled' towards the axes. In contrast, L2 regularization maintains a smooth, bowl-shaped landscape due to its quadratic penalty, thus favoring solutions near the origin without necessarily making any coefficients precisely zero.

5. A large learning rate α in a gradient descent algorithm can lead to overshooting the minima, particularly when L2 regularization is applied. Mathematically, the update $\mathtt{w}_j := \mathtt{w}_j - \alpha \left(\frac{\partial J}{\partial \mathtt{w}_j} + \lambda \mathtt{w}_j \right)$ means that large α can cause oscillations or instability as it increases the effect of weight shrinkage rapidly, potentially skipping optimal points or diverging the updates.

6. The regularization term $\|\mathbf{x}\|_1$ is non-differentiable at zero because the absolute value function $|x|$ has a kink at zero, where its derivative does not exist. Subgradient methods are used to handle this by employing subgradients, a generalization of gradients for non-differentiable points. At zero, any value between -1 and 1 is a subgradient, allowing the optimization algorithm to proceed in that range without precise derivatives.

Practice Problems 3

1. Explain the conceptual difference between L1 and L2 regularization in the context of a convex optimization problem. Provide a theoretical perspective on when each regularization might be preferred based on the geometry of the solution space.

2. Derive the gradient descent update rule for L1 regularization from the objective function and explain how it differs from L2 regularization in terms of handling sparsity.

3. Given a regularized cost function $J_{L1}(\mathbf{w}) = J(\mathbf{w}) + \lambda \sum_{j=1}^{n} |w_j|$, analyze the impact of varying λ on the solution's characteristics and discuss its implications on model complexity.

4. For the L2 regularization scenario, quantify how the choice of λ affects the eigenvalues of the Hessian matrix and infer its effect on the convergence properties of gradient-based optimization algorithms.

5. With $R(\mathbf{w}) = \|\mathbf{w}\|_2^2$, analyze the curvature of the regularization term and discuss its impact on ensuring a unique global minimum in the context of convex optimization problems.

6. Explain the geometric interpretation of L1 regularization leading to sparsity in the parameter vector, and discuss its advantages in high-dimensional data analysis.

Answers 3

1. **Conceptual Difference:** L1 regularization, or Lasso, adds an ℓ_1 penalty term to encourage sparsity, resulting in some weight coefficients being exactly zero. Geometrically, this corresponds to diamond-shaped contours which favor solutions aligned with the axes, thereby inducing sparsity. L2 regulariza-

tion, or Ridge, uses an ℓ_2 penalty, resulting in circular contours which uniformly shrink all coefficients. L1 is preferred for feature selection while L2 stabilizes solutions when all features contribute meaningfully to the outcome.

2. **Gradient Descent for L1:** Starting from $J_{L1}(\mathbf{w}) = J(\mathbf{w}) + \lambda \sum_{j=1}^{n} |w_j|$, the gradient descent update rule for L1 involves using the subgradient due to $|w_j|$ being non-differentiable at zero. At non-zero, $\text{sign}(w_j)$ is used in updates:

$$\mathbf{w}_j := \mathbf{w}_j - \alpha \left(\frac{\partial J}{\partial \mathbf{w}_j} + \lambda \cdot \text{sign}(w_j) \right)$$

In contrast, L2 straightforwardly adjusts weights proportionally to their size without requiring subgradient methods, leading to weight shrinkage without inducing zero weights.

3. **Impact of λ in L1:** As λ increases, the penalty for non-zero weights intensifies, promoting sparser solutions. Concretely, higher λ values force more weights to zero, enhancing simplicity but potentially omitting informative variables. Lower λ values provide flexibility in retaining contributive variables albeit with higher model complexity.

4. **Effect of λ on Hessian in L2:** The Hessian matrix in L2 regularization becomes $\nabla^2 J(\mathbf{w}) + \lambda I$, where I is the identity matrix. λ increases the minimum eigenvalue, improving conditioning of the cost function and convergence speed of gradient-based methods by smoothing cost surface curvature.

5. **Curvature in L2 Regularization:** The term $\|\mathbf{w}\|_2^2$ implies convexity due to its bowl-shaped parabola in function space, ensuring that any local minimum is a global minimum. The uniform penalization of coefficients strengthens convexity, contributing to straightforward convergence to a solution that balances complexity and approximation error.

6. **Geometric Interpretation of Sparsity in L1:** L1 regularization, by imposing an ℓ_1 penalty, results in contour lines that prefer crossing coordinate axes, leading to exact zeros (sparsity) in the solution. This is advantageous in high-dimensional settings, as it reduces model complexity and computation by retaining only the most informative features, minimizing overfitting while improving interpretability.

Chapter 22

Constrained Optimization

Practice Problems 1

1. Given the function $f(\mathbf{x}) = x^2 + y^2$ subject to the constraint $g(\mathbf{x}) = x + y - 1 = 0$, find the optimal solution using the method of Lagrange multipliers.

2. Prove that the Lagrangian method provides an optimal solution for minimizing $f(x, y) = x^2 + 4y^2$ with the constraint $x + 2y = 3$.

3. Determine the Lagrange multiplier λ when finding the maximum of $f(x, y) = 3x + 4y$ subject to $x^2 + y^2 = 10$.

4. Find the Lagrangian dual of the problem: Minimize $f(x) = x^2 - 2x$ subject to $g(x) = x - 1 = 0$.

5. Show that the solution to the primal problem of maximizing $f(x, y) = 2x + y$ with constraints $x^2 + y^2 \leq 4$ and $x \geq 0, y \geq 0$ can be solved using Lagrange multipliers.

6. For the function $f(x, y, z) = x + y + z$ with constraints $x^2 + y^2 + z^2 = 1$ and $x + y = 1$, find the values of x, y, z and the associated Lagrange multipliers, λ_1 and λ_2.

Answers 1

1. Given the function $f(\mathbf{x}) = x^2 + y^2$ subject to the constraint $g(\mathbf{x}) = x + y - 1 = 0$, find the optimal solution using the method of Lagrange multipliers.

   Solution:
   $$\mathcal{L}(x, y, \lambda) = x^2 + y^2 + \lambda(x + y - 1)$$

   Compute the gradients:
   $$\nabla \mathcal{L} = \left(\frac{\partial \mathcal{L}}{\partial x}, \frac{\partial \mathcal{L}}{\partial y}, \frac{\partial \mathcal{L}}{\partial \lambda} \right) = (2x + \lambda, 2y + \lambda, x + y - 1)$$

   Setting these equal to zero gives:
   $$2x + \lambda = 0,$$
   $$2y + \lambda = 0,$$
   $$x + y - 1 = 0.$$

   Solving these simultaneously yields $x = y = \frac{1}{2}$ and $\lambda = -1$.

2. Prove that the Lagrangian method provides an optimal solution for minimizing $f(x, y) = x^2 + 4y^2$ with the constraint $x + 2y = 3$.

   Solution:
   $$\mathcal{L}(x, y, \lambda) = x^2 + 4y^2 + \lambda(x + 2y - 3)$$

   Compute the gradients:
   $$\nabla \mathcal{L} = (2x + \lambda, 8y + 2\lambda, x + 2y - 3)$$

   Set to zero:
   $$2x + \lambda = 0,$$
   $$8y + 2\lambda = 0,$$
   $$x + 2y - 3 = 0.$$

216

Solve the first equation for λ: $\lambda = -2x$. Substitute into the second: $8y = -4x$ gives $y = -\frac{x}{2}$. Substitute $y = -\frac{x}{2}$ into the constraint: $x - x = 3$. Solving the system, we find $x = 1$, $y = 1$, and $\lambda = -2$.

3. Determine the Lagrange multiplier λ when finding the maximum of $f(x, y) = 3x + 4y$ subject to $x^2 + y^2 = 10$.

   Solution:

$$\mathcal{L}(x, y, \lambda) = 3x + 4y + \lambda(10 - x^2 - y^2)$$

   Compute the gradients:

$$\nabla \mathcal{L} = (3 - 2\lambda x, 4 - 2\lambda y, 10 - x^2 - y^2)$$

   Set the gradients to zero:

$$3 - 2\lambda x = 0,$$
$$4 - 2\lambda y = 0,$$
$$x^2 + y^2 = 10.$$

   Solving the first, $\lambda = \frac{3}{2x}$. From the second, $\lambda = \frac{2}{y}$. Equating gives $\frac{3}{2x} = \frac{2}{y}$, solving gives the multiplier $\lambda = \frac{1.5}{x} = \frac{2}{y}$.

4. Find the Lagrangian dual of the problem: Minimize $f(x) = x^2 - 2x$ subject to $g(x) = x - 1 = 0$.

   Solution:

$$\mathcal{L}(x, \lambda) = x^2 - 2x + \lambda(x - 1)$$

   Compute the dual by setting:

$$\inf_x \mathcal{L}(x, \lambda) = x^2 + (\lambda - 2)x - \lambda$$

   Complete the square to find:

$$\inf_x = \frac{(2 - \lambda)^2}{4} - \lambda$$

   Therefore, the dual function $d(\lambda)$ is:

$$d(\lambda) = -\frac{1}{4}(\lambda^2 - 4\lambda + 4) - \lambda = -\frac{1}{4}\lambda^2 + \frac{\lambda}{2} - 1$$

5. Show that the solution to the primal problem of maximizing $f(x, y) = 2x + y$ with constraints $x^2 + y^2 \leq 4$ and $x \geq 0, y \geq 0$ can be solved using Lagrange multipliers.

   Solution: Introduce the constraint $g(x, y) = x^2 + y^2 - 4 \leq 0$. Using a slack variable, the active constraint simplifies calculation:

$$\mathcal{L}(x, y, \lambda) = 2x + y + \lambda(4 - x^2 - y^2)$$

   Calculate gradients:

$$\nabla \mathcal{L} = \left(2 - 2\lambda x, 1 - 2\lambda y, 4 - x^2 - y^2\right)$$

   Set gradients equal to zero:

$$2 - 2\lambda x = 0 \Rightarrow \lambda = \frac{1}{x},$$
$$1 - 2\lambda y = 0 \Rightarrow \lambda = \frac{1}{2y},$$
$$x^2 + y^2 = 4.$$

   Equating the two expressions for λ results in $x = 2y$. Substituting into the constraint: $(2y)^2 + y^2 = 4$ gives $5y^2 = 4$, solving gives $y = \sqrt{\frac{4}{5}}$.

6. For the function $f(x, y, z) = x + y + z$ with constraints $x^2 + y^2 + z^2 = 1$ and $x + y = 1$, find the values of x, y, z and the associated Lagrange multipliers, λ_1 and λ_2.

Solution:

$$\mathcal{L}(x, y, z, \lambda_1, \lambda_2) = x + y + z + \lambda_1(1 - x^2 - y^2 - z^2) + \lambda_2(1 - x - y)$$

Compute the gradients:

$$\nabla \mathcal{L} = \left(1 - 2\lambda_1 x - \lambda_2, 1 - 2\lambda_1 y - \lambda_2, 1 - 2\lambda_1 z, 1 - x^2 - y^2 - z^2, x + y - 1\right)$$

Set to zero:

$$1 - 2\lambda_1 x - \lambda_2 = 0,$$
$$1 - 2\lambda_1 y - \lambda_2 = 0,$$
$$1 - 2\lambda_1 z = 0,$$
$$x^2 + y^2 + z^2 = 1,$$
$$x + y = 1.$$

Solving this set, using symmetry and constraints:

$$\lambda_1 = \frac{1}{2x} = \frac{1}{2y} = \frac{1}{2z}, \quad x = y = z = \frac{\sqrt{1/3}}{\sqrt{2}}, \quad \lambda_2 = 0.$$

Practice Problems 2

1. Formulate the Lagrangian for the problem of minimizing $h(x, y) = x^2 + y^2$ subject to the constraint $x + 2y = 3$. Find the stationary points of the Lagrangian.

2. For the function $k(x, y) = x^2 - y$ with the constraint $x^2 + y^2 = 1$, set up the Lagrangian and determine the critical points.

3. Consider the function $p(x, y, z) = xyz$ subject to $x^2 + y^2 + z^2 = 1$. Construct the Lagrangian and solve for the maximum values of p.

4. Solve the optimization problem of maximizing $f(x, y) = xy$ given the constraint $2x + y = 10$.

5. Using Lagrange multipliers, determine the points that maximize $m(x, y) = x^2 + 4y^2$ under the constraint $x + y = 5$.

6. For the function $s(x, y) = 3x + 4y$ with the constraint $x^2 + y^2 = 25$, find the minimum using Lagrangian multipliers.

Answers 2

1. **Solution:**

$$\text{Lagrangian } \mathcal{L}(x, y, \lambda) = x^2 + y^2 + \lambda(x + 2y - 3)$$

Compute the gradients:

$$\frac{\partial \mathcal{L}}{\partial x} = 2x + \lambda, \quad \frac{\partial \mathcal{L}}{\partial y} = 2y + 2\lambda, \quad \frac{\partial \mathcal{L}}{\partial \lambda} = x + 2y - 3$$

Set each derivative to zero to find critical points:

$$2x + \lambda = 0, \quad 2y + 2\lambda = 0, \quad x + 2y - 3 = 0$$

Solving these, $x = 1, y = 1, \lambda = -2$. Thus, the stationary point is $(1, 1, -2)$.

2. **Solution:**

$$\text{Lagrangian } \mathcal{L}(x, y, \lambda) = x^2 - y + \lambda(x^2 + y^2 - 1)$$

Compute the gradients:

$$\frac{\partial \mathcal{L}}{\partial x} = 2x + 2\lambda x, \quad \frac{\partial \mathcal{L}}{\partial y} = -1 + 2\lambda y, \quad \frac{\partial \mathcal{L}}{\partial \lambda} = x^2 + y^2 - 1$$

Set each derivative to zero:

$$2x + 2\lambda x = 0, \quad -1 + 2\lambda y = 0, \quad x^2 + y^2 - 1 = 0$$

From $2x(1 + \lambda) = 0$, if $x \neq 0$, $\lambda = -1$. From $-1 + 2\lambda y = 0$, $y = \frac{1}{2\lambda}$, solving gives $y = -\frac{1}{2}$ and constraint gives $x = \sqrt{\frac{3}{4}}$.

3. **Solution:**

$$\text{Lagrangian } \mathcal{L}(x, y, z, \lambda) = xyz + \lambda(x^2 + y^2 + z^2 - 1)$$

Compute the gradients:

$$\frac{\partial \mathcal{L}}{\partial x} = yz + 2\lambda x, \quad \frac{\partial \mathcal{L}}{\partial y} = xz + 2\lambda y, \quad \frac{\partial \mathcal{L}}{\partial z} = xy + 2\lambda z$$

$$\frac{\partial \mathcal{L}}{\partial \lambda} = x^2 + y^2 + z^2 - 1$$

Set derivatives to zero gives system:

$$yz + 2\lambda x = 0, \quad xz + 2\lambda y = 0, \quad xy + 2\lambda z = 0$$

$$x^2 + y^2 + z^2 = 1$$

Assume $\lambda = 0$, gives points on the sphere $(x, y, z) = (1/\sqrt{3}, 1/\sqrt{3}, 1/\sqrt{3})$.

4. **Solution:**

$$\text{Lagrangian } \mathcal{L}(x, y, \lambda) = xy + \lambda(2x + y - 10)$$

Compute the gradients:

$$\frac{\partial \mathcal{L}}{\partial x} = y + 2\lambda, \quad \frac{\partial \mathcal{L}}{\partial y} = x + \lambda$$

Set each derivative to zero:

$$y + 2\lambda = 0, \quad x + \lambda = 0, \quad 2x + y = 10$$

Solving gives $x = 4, y = 2, \lambda = -4$, so maximum is at $(4, 2)$.

5. **Solution:**

$$\mathcal{L}(x, y, \lambda) = x^2 + 4y^2 + \lambda(x + y - 5)$$

Calculate gradients:

$$\frac{\partial \mathcal{L}}{\partial x} = 2x + \lambda, \quad \frac{\partial \mathcal{L}}{\partial y} = 8y + \lambda$$

Derivatives to zero:

$$2x + \lambda = 0, \quad 8y + \lambda = 0, \quad x + y = 5$$

Solving to find $x = 5, y = 0, \lambda = -10$.

6. **Solution:**

$$\text{Lagrangian } \mathcal{L}(x, y, \lambda) = 3x + 4y + \lambda(x^2 + y^2 - 25)$$

Compute the gradients:

$$\frac{\partial \mathcal{L}}{\partial x} = 3 + 2\lambda x, \quad \frac{\partial \mathcal{L}}{\partial y} = 4 + 2\lambda y$$

Set derivatives to zero:

$$3 + 2\lambda x = 0, \quad 4 + 2\lambda y = 0, \quad x^2 + y^2 = 25$$

Solving yields points: $x = -\frac{3}{2\lambda}, y = -\frac{2}{\lambda}$, using the constraint gives $\lambda = \pm \frac{5}{\sqrt{13}}$, with specific values leading to points on the circle.

Practice Problems 3

1. Consider a function $f(x, y) = x^2 + y^2$ and the constraint $g(x, y) = x + y - 1 = 0$. Use Lagrange multipliers to find the extreme values of f.

2. Show that the critical points found using the Lagrange multiplier method for a function $f(x, y) = 2x + y$ under the constraint $g(x, y) = x^2 + y^2 - 4 = 0$ are indeed extrema.

3. Derive the dual problem from the primal problem of maximizing $f(x, y) = 3x + 5y$ subject to $2x + y \leq 4$ and $x, y \geq 0$. Assume x, y and the dual variables are non-negative.

4. In a constrained optimization problem, verify the convexity of the primal problem: minimize $f(x) = x_1^2 + x_2^2$ subject to $x_1 + 2x_2 = 4$. Use the properties of convex sets.

5. Explain the role of the constraint qualification condition (such as the Slater's condition) in ensuring strong duality for the optimization problem $\min f(x) = x^2$ subject to $g(x) = x - 1 \leq 0$.

6. For a constrained optimization problem of maximizing $f(x,y) = 4x + y$ with the constraint $g(x,y) = x^2 + y^2 \leq 1$, determine if the optimal solution satisfies the Karush-Kuhn-Tucker (KKT) conditions.

Answers 3

1. Consider a function $f(x,y) = x^2 + y^2$ and the constraint $g(x,y) = x + y - 1 = 0$.

   Solution:

   Construct the Lagrangian:
   $$\mathcal{L}(x,y,\lambda) = x^2 + y^2 + \lambda(x + y - 1)$$

   Compute the gradients and set them to zero:
   $$\frac{\partial \mathcal{L}}{\partial x} = 2x + \lambda = 0, \quad \frac{\partial \mathcal{L}}{\partial y} = 2y + \lambda = 0, \quad \frac{\partial \mathcal{L}}{\partial \lambda} = x + y - 1 = 0$$

   Solve the system of equations:
   $$2x + \lambda = 0 \quad (1)$$
   $$2y + \lambda = 0 \quad (2)$$
   $$x + y - 1 = 0 \quad (3)$$

   From equations (1) and (2), $2x = 2y$ leads to $x = y$. Substituting in (3) gives $x + x - 1 = 0 \Rightarrow x = \frac{1}{2}$ and $y = \frac{1}{2}$. Therefore, the extreme values are at $\left(\frac{1}{2}, \frac{1}{2}\right)$.

2. Show that the critical points found using the Lagrange multiplier method are extrema.

   Solution:

   Given $f(x,y) = 2x + y$ and constraint $g(x,y) = x^2 + y^2 - 4 = 0$, Lagrangian:
   $$\mathcal{L}(x,y,\lambda) = 2x + y + \lambda(x^2 + y^2 - 4)$$

   Gradients:
   $$\frac{\partial \mathcal{L}}{\partial x} = 2 + 2\lambda x = 0 \quad (1)$$
   $$\frac{\partial \mathcal{L}}{\partial y} = 1 + 2\lambda y = 0 \quad (2)$$
   $$\frac{\partial \mathcal{L}}{\partial \lambda} = x^2 + y^2 - 4 = 0 \quad (3)$$

   Solving these, $\lambda = -\frac{1}{x} = -\frac{1}{2y}$. Solving for x and y using (3), generally leads to: $x^2 + \left(\frac{x}{2}\right)^2 = 4$. This verified $x = 2$ and $y = -1$, showing valid extrema by constraint substitution.

3. Derive dual problem.

   Solution:

   Primal: Maximize $3x + 5y$ subject to $2x + y \leq 4$, $x, y \geq 0$. Lagrangian with slack variable $s \geq 0$:
   $$\mathcal{L}(x,y,s,\lambda_1,\lambda_2,\lambda_3) = 3x + 5y + \lambda_1(4 - 2x - y) + \lambda_2 x + \lambda_3 y$$

Karush-Kuhn-Tucker conditions lead to:

$$\lambda_1, \lambda_2, \lambda_3 \geq 0; \quad 2\lambda_1 - \lambda_2 = 3; \quad \lambda_1 - \lambda_3 = 5$$

From leveraging primal slackness:

$$\lambda_1(2x + y - 4) = 0, \quad \lambda_2 x = 0, \quad \lambda_3 y = 0$$

Dual problem construction seeks to maximize the independent components directly based on associated multiplier strengths.

4. Verify convexity with properties.

   Solution:

   Function $f(x_1, x_2) = x_1^2 + x_2^2$ is quadratic and convex as Hessian matrix is positive semi-definite:

$$H = \begin{bmatrix} 2 & 0 \\ 0 & 2 \end{bmatrix}$$

   Constraint set $x_1 + 2x_2 = 4$ represents linear equality, encloses convex region leading to overall problem convex for minimized objective.

5. Role of constraint qualification in duality.

   Solution:

   For $\min f(x) = x^2$ subject to $g(x) = x - 1 \leq 0$, verifying Slater's condition $g(x) < 0$ is pivotal: Constraint simplifies directly since $x < 1$ allows distinct boundary compliance. Strong duality is achieved, building correspondence between primal-optimal point and this enhanced constraint state underlying same-point duality; hence critical for equivalence validation.

6. Determine KKT condition satisfication.

   Solution:

   Objective $f(x, y) = 4x + y$ with $g(x, y) = x^2 + y^2 \leq 1$. KKT constraints lead:

$$L(x, y, \lambda) = 4x + y + \lambda(x^2 + y^2 - 1)$$

   Gradients:

$$\frac{\partial L}{\partial x} = 4 + 2\lambda x = 0, \quad \frac{\partial L}{\partial y} = 1 + 2\lambda y = 0$$

   Confirm by feasibility: solutions within $x^2 + y^2 \leq 1$ and adjusted multiplier λ yield internal saddle point; proving KKT satisfaction.

Chapter 23

Probability Density Functions

Practice Problems 1

1. Evaluate the probability that a continuous random variable X with a normal distribution $N(\mu, \sigma^2)$ falls within one standard deviation of the mean.

2. For a random variable X with an exponential distribution having rate parameter λ, calculate the probability $P(X > x)$.

3. Consider a uniform distribution defined over $[a, b]$. Derive the expression for the expected value $E(X)$ for a random variable X.

4. Prove that the normal distribution is symmetrical about its mean μ.

5. For a standard normal distribution, calculate the probability $P(-1 \leq Z \leq 1)$ using the properties of the standard normal distribution.

6. Derive the variance of an exponential distribution with the rate parameter λ.

Answers 1

1. Evaluate the probability that a continuous random variable X with a normal distribution $N(\mu, \sigma^2)$ falls within one standard deviation of the mean.

   Solution: For $X \sim N(\mu, \sigma^2)$, we want to find $P(\mu - \sigma \leq X \leq \mu + \sigma)$. By the empirical rule (68–95–99.7 rule), we know:

   $$P(\mu - \sigma \leq X \leq \mu + \sigma) \approx 0.68$$

   Therefore, the probability is approximately 0.68.

2. For a random variable X with an exponential distribution having rate parameter λ, calculate the probability $P(X > x)$.

   Solution: The cumulative distribution function (CDF) of an exponential distribution is:

   $$F(x) = 1 - e^{-\lambda x}, \quad x \geq 0$$

   Hence, the probability $P(X > x) = 1 - F(x)$:

   $$P(X > x) = e^{-\lambda x}$$

3. Consider a uniform distribution defined over $[a, b]$. Derive the expression for the expected value $E(X)$ for a random variable X.

   Solution: The PDF of a uniform distribution is:

   $$f(x) = \frac{1}{b - a}, \quad a \leq x \leq b$$

   The expected value is calculated as:

   $$E(X) = \int_a^b x f(x)\, dx = \int_a^b x \frac{1}{b - a}\, dx$$

Solving the integral:

$$= \frac{1}{b-a} \left[\frac{x^2}{2} \right]_a^b = \frac{1}{b-a} \left(\frac{b^2}{2} - \frac{a^2}{2} \right)$$

$$= \frac{b^2 - a^2}{2(b-a)} = \frac{(b+a)(b-a)}{2(b-a)}$$

$$= \frac{b+a}{2}$$

Therefore, $E(X) = \frac{b+a}{2}$.

4. Prove that the normal distribution is symmetrical about its mean μ.

Solution: The PDF of a normal distribution is:

$$f(x \mid \mu, \sigma) = \frac{1}{\sigma\sqrt{2\pi}} e^{-\frac{(x-\mu)^2}{2\sigma^2}}$$

For symmetry, show $f(\mu + d) = f(\mu - d)$ for any d. Substitute $x = \mu + d$ and $x = \mu - d$ into the PDF:

$$f(\mu + d) = \frac{1}{\sigma\sqrt{2\pi}} e^{-\frac{d^2}{2\sigma^2}}$$

$$f(\mu - d) = \frac{1}{\sigma\sqrt{2\pi}} e^{-\frac{d^2}{2\sigma^2}}$$

Since both expressions are equal, the normal distribution is symmetric around μ.

5. For a standard normal distribution, calculate the probability $P(-1 \leq Z \leq 1)$ using the properties of the standard normal distribution.

Solution: For $Z \sim N(0, 1)$, we use the empirical rule:

$$P(-1 \leq Z \leq 1) \approx 0.68$$

This probability approximates to 68

6. Derive the variance of an exponential distribution with the rate parameter λ.

Solution: The mean of the exponential distribution is:

$$E(X) = \frac{1}{\lambda}$$

The variance is determined by calculating $E(X^2)$ and using $\text{Var}(X) = E(X^2) - (E(X))^2$.

$$E(X^2) = \int_0^\infty x^2 \lambda e^{-\lambda x} \, dx$$

Integration by parts yields:

$$= \frac{2}{\lambda^2}$$

Thus, the variance is:

$$\text{Var}(X) = \frac{2}{\lambda^2} - \left(\frac{1}{\lambda} \right)^2 = \frac{1}{\lambda^2}$$

Therefore, the variance is $\frac{1}{\lambda^2}$.

Practice Problems 2

1. Given a continuous random variable X with probability density function $f(x) = \frac{1}{4}e^{-x/4}$ for $x \geq 0$, find the probability that X is between 2 and 5.

2. Consider a continuous random variable with a uniform distribution over the interval $[0, 10]$. Calculate the probability that a realization of this random variable is less than 3.

3. The probability density function of a random variable X is given as $f(x) = \frac{3}{8}(2x - x^2)$ for $0 \leq x \leq 2$. Verify that $f(x)$ is a valid probability density function.

4. For a normally distributed random variable with mean $\mu = 0$ and standard deviation $\sigma = 1$, compute the proportion of observations less than $z = 1.96$.

5. Given a random variable X with PDF $f(x) = \begin{cases} 0, & x < 0 \\ \frac{1}{2}, & 0 \leq x \leq 2 \\ 0, & x > 2 \end{cases}$, calculate $E(X)$.

6. Determine the variance of a random variable X distributed according to the exponential distribution with rate parameter $\lambda = 3$.

Answers 2

1. **Solution:** Evaluate the probability for X between 2 and 5:

$$P(2 \leq X \leq 5) = \int_2^5 \frac{1}{4} e^{-x/4} \, dx$$

$$= \left[-e^{-x/4} \right]_2^5$$

$$= -e^{-5/4} + e^{-2/4}$$

$$= e^{-1/2} - e^{-5/4}$$

Thus, the probability is approximately 0.1935.

2. **Solution:** For a uniform distribution $U(0, 10)$:

$$P(X < 3) = \frac{3 - 0}{10 - 0} = \frac{3}{10} = 0.3$$

Hence, $P(X < 3) = 0.3$.

3. **Solution:** Check normalization of $f(x)$:

$$\int_0^2 \frac{3}{8} (2x - x^2) \, dx$$

$$= \frac{3}{8} \left[x^2 - \frac{x^3}{3} \right]_0^2$$

$$= \frac{3}{8} \left(4 - \frac{8}{3} \right)$$

$$= \frac{3}{8} \cdot \frac{4}{3} = 1$$

Therefore, $f(x)$ is a valid PDF.

4. **Solution:** Use standard normal table for $z = 1.96$:

$$P(Z < 1.96) \approx 0.975$$

Thus, the proportion is approximately 0.975.

5. **Solution:** Calculate $E(X)$ for the given PDF:

$$E(X) = \int_0^2 x \cdot \frac{1}{2} \, dx$$

$$= \frac{1}{2} \left[\frac{x^2}{2} \right]_0^2$$

$$= \frac{1}{2} \cdot 2 = 1$$

Hence, $E(X) = 1$.

6. **Solution:** Find variance of exponential $\lambda = 3$:

$$\text{Var}(X) = \frac{1}{\lambda^2} = \frac{1}{9}$$

Therefore, the variance is $\frac{1}{9}$.

Practice Problems 3

1. Consider a continuous random variable X with probability density function $f(x)$ given by:

$$f(x) = \begin{cases} \frac{1}{2} e^{-\frac{x}{2}} & \text{for } x \geq 0 \\ 0 & \text{otherwise} \end{cases}$$

Verify that this is a valid probability density function.

2. Calculate the probability that a random variable X with the probability density function from Problem 1 falls between 1 and 3.

3. Given a normal random variable $X \sim N(0, 1)$, find the probability that X is less than 1.

229

4. A random variable X follows a uniform distribution over the interval $[0, 5]$. Determine the probability density function for X and calculate the probability that X is less than 2.

5. If $X \sim N(\mu, \sigma^2)$, show that the mode of the distribution occurs at $X = \mu$.

6. Compute the expected value of a random variable X with an exponential distribution defined by the following PDF:
$$f(x) = \lambda e^{-\lambda x}, \quad \text{for } x \geq 0$$

where $\lambda > 0$.

Answers 3

1. **Solution:** Verify that $f(x) = \frac{1}{2}e^{-\frac{x}{2}}$ for $x \geq 0$ is a valid probability density function:

$$\int_0^\infty \frac{1}{2}e^{-\frac{x}{2}}\, dx$$

Let $u = -\frac{x}{2}$ then $du = -\frac{1}{2}\, dx$ thus $dx = -2\, du$.

$$= \int_{-\infty}^0 \frac{1}{2}e^u \cdot (-2)\, du$$

$$= \int_{-\infty}^0 e^u\, du$$

$$= [e^u]_{-\infty}^0$$

$$= 1 - 0 = 1$$

Therefore, $f(x)$ is a valid PDF.

2. **Solution:** Calculate the probability that X with PDF as given in Problem 1 falls between 1 and 3:

$$P(1 \leq X \leq 3) = \int_1^3 \frac{1}{2} e^{-\frac{x}{2}} \, dx$$

$$= \left[-e^{-\frac{x}{2}} \right]_1^3$$

$$= -e^{-\frac{3}{2}} - (-e^{-\frac{1}{2}})$$

$$= e^{-\frac{1}{2}} - e^{-\frac{3}{2}}$$

$$\approx 0.303 - 0.112 \approx 0.191$$

Therefore, the probability is approximately 0.191.

3. **Solution:** Find $P(X < 1)$ if $X \sim N(0, 1)$:

$$P(X < 1) = \Phi(1)$$

Using standard normal distribution tables or software,

$$\Phi(1) \approx 0.8413$$

Therefore, the probability is approximately 0.8413.

4. **Solution:** For $X \sim U(0, 5)$:

$$f(x) = \frac{1}{b-a} = \frac{1}{5-0} = \frac{1}{5}, \quad 0 \leq x \leq 5$$

Calculate $P(X < 2)$:

$$P(X < 2) = \int_0^2 \frac{1}{5} \, dx$$

$$= \frac{1}{5} \cdot [x]_0^2$$

$$= \frac{1}{5} \cdot 2 = \frac{2}{5} = 0.4$$

Therefore, the probability is 0.4.

5. **Solution:** Show mode for $X \sim N(\mu, \sigma^2)$ is at $X = \mu$:

$$f(x) = \frac{1}{\sigma\sqrt{2\pi}} e^{-\frac{(x-\mu)^2}{2\sigma^2}}$$

The function achieves its maximum where the exponent is zero:

$$-\frac{(x-\mu)^2}{2\sigma^2} \quad \text{maximized when } (x-\mu)^2 = 0$$

$$\Rightarrow x = \mu$$

Thus the mode is μ.

6. **Solution:** Compute expected value for X with PDF $f(x) = \lambda e^{-\lambda x}$:

$$E[X] = \int_0^\infty x \cdot \lambda e^{-\lambda x}\, dx$$

By integration by parts, let $u = x$, $dv = \lambda e^{-\lambda x}\, dx$:

$$du = dx, \quad v = -e^{-\lambda x}$$

$$E[X] = \left[-xe^{-\lambda x} \right]_0^\infty + \int_0^\infty e^{-\lambda x}\, dx$$

Evaluating,

$$= \left[\frac{-x}{e^{\lambda x}} \right]_0^\infty + \left[\frac{-1}{\lambda} e^{-\lambda x} \right]_0^\infty$$

$$= 0 - 0 + \frac{1}{\lambda}$$

Therefore, $E[X] = \frac{1}{\lambda}$.

Chapter 24

Cost Functions in Machine Learning

Practice Problems 1

1. Consider the cost function defined by the Mean Squared Error (MSE):

$$J(\theta) = \frac{1}{m} \sum_{i=1}^{m} (h_\theta(x_i) - y_i)^2$$

Compute $\frac{\partial J(\theta)}{\partial \theta_j}$.

2. For binary classification, the cross-entropy loss is defined as:

$$J(\theta) = -\frac{1}{m} \sum_{i=1}^{m} (y_i \log(h_\theta(x_i)) + (1 - y_i) \log(1 - h_\theta(x_i)))$$

Derive the gradient $\frac{\partial J(\theta)}{\partial \theta_j}$.

3. For the gradient descent algorithm, show how the update rule:

$$\theta_j := \theta_j - \alpha \frac{\partial J(\theta)}{\partial \theta_j}$$

reduces the cost function $J(\theta)$ at each iteration.

4. Given the quadratic cost function:

$$J(\theta) = \frac{1}{2m} \sum_{i=1}^{m} (h_\theta(x_i) - y_i)^2$$

Find the closed-form solution for θ using the normal equation.

5. Consider the regularized cost function:

$$J(\theta) = \frac{1}{m} \sum_{i=1}^{m} (h_\theta(x_i) - y_i)^2 + \lambda \sum_{j=1}^{n} \theta_j^2$$

Compute the gradient with respect to θ_j.

6. Analyze how the curvature information provided by the Hessian matrix:

$$H = \nabla^2 J(\theta)$$

can be utilized in Newton's method to achieve faster convergence compared to gradient descent alone.

Answers 1

1. Compute $\frac{\partial J(\theta)}{\partial \theta_j}$ for the Mean Squared Error.

   Solution:

   $$J(\theta) = \frac{1}{m} \sum_{i=1}^{m} (h_\theta(x_i) - y_i)^2$$

   Differentiating with respect to θ_j, we use the chain rule:

   $$\frac{\partial J(\theta)}{\partial \theta_j} = \frac{1}{m} \sum_{i=1}^{m} 2(h_\theta(x_i) - y_i) \frac{\partial h_\theta(x_i)}{\partial \theta_j}$$

   $$= \frac{2}{m} \sum_{i=1}^{m} (h_\theta(x_i) - y_i) x_{ij}$$

   Hence,

   $$\frac{\partial J(\theta)}{\partial \theta_j} = \frac{2}{m} \sum_{i=1}^{m} (h_\theta(x_i) - y_i) x_{ij}.$$

2. Derive the gradient $\frac{\partial J(\theta)}{\partial \theta_j}$ for the cross-entropy loss.

   Solution:

   $$J(\theta) = -\frac{1}{m} \sum_{i=1}^{m} (y_i \log(h_\theta(x_i)) + (1 - y_i) \log(1 - h_\theta(x_i)))$$

   Differentiating:

   $$\frac{\partial J(\theta)}{\partial \theta_j} = -\frac{1}{m} \sum_{i=1}^{m} \left(\frac{y_i}{h_\theta(x_i)} \cdot \frac{\partial h_\theta(x_i)}{\partial \theta_j} - \frac{1 - y_i}{1 - h_\theta(x_i)} \cdot \frac{\partial h_\theta(x_i)}{\partial \theta_j} \right)$$

   Simplifying yields:

   $$= \frac{1}{m} \sum_{i=1}^{m} ((h_\theta(x_i) - y_i) x_{ij})$$

   Therefore, the gradient for the cross-entropy loss is:

   $$\frac{\partial J(\theta)}{\partial \theta_j} = \frac{1}{m} \sum_{i=1}^{m} (h_\theta(x_i) - y_i) x_{ij}.$$

3. Show how the update rule reduces the cost function in gradient descent.

   Solution: The update rule is:

   $$\theta_j := \theta_j - \alpha \frac{\partial J(\theta)}{\partial \theta_j}$$

   From Taylor's theorem, $J(\theta - \alpha \nabla J(\theta)) \approx J(\theta) - \alpha \|\nabla J(\theta)\|^2$.

   Since α and $\|\nabla J(\theta)\|^2$ are non-negative, each update reduces $J(\theta)$, ensuring decrease unless $\nabla J(\theta) = 0$.

4. Find the closed-form solution using the normal equation for the quadratic cost.

   Solution: The cost function is:

   $$J(\theta) = \frac{1}{2m} \sum_{i=1}^{m} (h_\theta(x_i) - y_i)^2$$

   For linear regression, $h_\theta(x_i) = \theta^T x_i$.

   Normal equation:

   $$\theta = (X^T X)^{-1} X^T \mathbf{y}$$

   where X is the feature matrix and $\mathbf{y}$ is the output vector, providing direct minimization of $J(\theta)$.

5. Compute the gradient with regularization.

   Solution:

   $$J(\theta) = \frac{1}{m} \sum_{i=1}^{m} (h_\theta(x_i) - y_i)^2 + \lambda \sum_{j=1}^{n} \theta_j^2$$

   Compute:

   $$\frac{\partial J(\theta)}{\partial \theta_j} = \frac{2}{m} \sum_{i=1}^{m} (h_\theta(x_i) - y_i)x_{ij} + 2\lambda\theta_j$$

   Thus, regularization adds $2\lambda\theta_j$ to the gradient, controlling weight magnitude.

6. Utilize the Hessian for faster convergence in Newton's method.

   Solution: Newton's update:

   $$\theta := \theta - H^{-1}\nabla J(\theta)$$

   where $H = \nabla^2 J(\theta)$.

   The Hessian H provides curvature (second derivative) info; this helps account for changes in the gradient.

   By considering the Hessian, Newton's method adjusts steps according to the terrain of $J(\theta)$, achieving quadratic convergence in ideal conditions.

   This curvature-awareness adjusts the leap towards minima compared to gradient descent, which solely uses the slope (first derivative) and can stagnate on plateaus or overshoot.

Practice Problems 2

1. For the dataset $\{(x_i, y_i)\}$, demonstrate how the Mean Squared Error (MSE) cost function is derived and explain its significance in linear regression.

2. Given the cross-entropy loss function for a binary classification problem, derive the gradient of this loss with respect to the model parameters θ.

3. Explain the convergence properties of the gradient descent algorithm when applied to a convex quadratic cost function.

4. Describe the role of the learning rate α in the gradient descent algorithm and its impact on convergence. Also, discuss the consequences of choosing α too small or too large.

5. Assume a linear regression model $h_\theta(x) = \theta_0 + \theta_1 x$ and derive the normal equation to find the optimal parameters.

6. For a cost function with both convex and non-convex components, analyze the challenges posed in optimization and how techniques like momentum could alleviate these challenges.

Answers 2

1. **Derivation of the Mean Squared Error (MSE) Cost Function:**

   Solution: The Mean Squared Error (MSE) quantifies the average of the squares of the errors, where the error is the difference between the predicted values $h_\theta(x_i)$ and the actual values y_i. The cost function is derived as follows:

   $$J(\theta) = \frac{1}{m} \sum_{i=1}^{m} (h_\theta(x_i) - y_i)^2$$

   Significance: MSE is used to evaluate the performance of a linear regression model, emphasizing larger errors by squaring the differences, which can be beneficial in model training for minimizing these errors.

2. **Deriving the Gradient of Cross-Entropy Loss:**

   Solution: For cross-entropy loss in binary classification:

   $$J(\theta) = -\frac{1}{m} \sum_{i=1}^{m} (y_i \log(h_\theta(x_i)) + (1 - y_i) \log(1 - h_\theta(x_i)))$$

The gradient w.r.t θ is:

$$\frac{\partial J(\theta)}{\partial \theta_j} = -\frac{1}{m} \sum_{i=1}^{m} \left(\frac{y_i}{h_\theta(x_i)} \frac{\partial h_\theta(x_i)}{\partial \theta_j} - \frac{1 - y_i}{1 - h_\theta(x_i)} \frac{\partial h_\theta(x_i)}{\partial \theta_j} \right)$$

This gradient guides the optimization process by indicating the direction to adjust θ for reducing the cost.

3. **Convergence of Gradient Descent on Convex Functions:**

   Solution: For a convex quadratic cost function, characterized as $J(\theta) = \frac{1}{2}\theta^T A\theta - b^T\theta$, gradient descent converges to the global minimum, as convex functions have a single minimum. The rate of convergence depends on the conditioning of A; well-conditioned matrices demonstrate faster convergence.

4. **Role of Learning Rate in Gradient Descent:**

   Solution: The learning rate α determines the step size during each iteration of gradient descent. A too-small α results in slow convergence, potentially getting stuck in flat areas; a too-large α can cause overshooting, leading to divergence. Finding the balance is crucial for efficient convergence.

5. **Normal Equation for Linear Regression:**

   Solution: For a linear model $h_\theta(x) = \theta_0 + \theta_1 x$, the normal equation minimizes the cost without iteration:
   $$\theta = (X^T X)^{-1} X^T \mathbf{y}$$

   Derivation: Assume an input matrix X and target vector $\mathbf{y}$, solve $(X^T X)\theta = X^T \mathbf{y}$ for θ.

6. **Optimization Challenges with Mixed Convexity:**

   Solution: Cost functions with both convex and non-convex parts present issues like local minima. Momentum, by considering historical gradients, helps escape such minima by smoothing the path and accelerating towards convergence, addressing oscillations and improving the convergence rate.

Practice Problems 3

1. Given the cost function defined by the Mean Squared Error (MSE) as $J(\theta) = \frac{1}{m} \sum_{i=1}^{m} (h_\theta(x_i) - y_i)^2$, find the derivative of $J(\theta)$ with respect to θ.

2. Consider the cross-entropy loss function for logistic regression
   $J(\theta) = -\frac{1}{m} \sum_{i=1}^{m} (y_i \log(h_\theta(x_i)) + (1 - y_i) \log(1 - h_\theta(x_i)))$. Determine $\frac{\partial J}{\partial \theta_j}$ for the model prediction $h_\theta(x_i) = \frac{1}{1 + e^{-\theta^T x_i}}$.

3. If the cost function includes a regularization term such as $J(\theta) = \frac{1}{m} \sum_{i=1}^{m} (h_\theta(x_i) - y_i)^2 + \lambda \sum_{j=1}^{n} \theta_j^2$, compute the gradient $\nabla J(\theta)$.

4. For a linear regression model, derive the closed-form solution using the normal equation given the matrix of inputs X and target vector $\mathbf{y}$.

5. Analyze the gradient descent update rule $\theta := \theta - \alpha \nabla J(\theta)$ and discuss its convergence criteria.

6. Define the impact of the Hessian matrix on second-order optimization methods such as Newton's method and its significance in the optimization of cost functions.

Answers 3

1. **Solution:**

The derivative of $J(\theta) = \frac{1}{m} \sum_{i=1}^{m} (h_\theta(x_i) - y_i)^2$ with respect to θ is:

$$\nabla J(\theta) = \frac{1}{m} \sum_{i=1}^{m} 2(h_\theta(x_i) - y_i) \frac{\partial h_\theta(x_i)}{\partial \theta}$$

Assuming $h_\theta(x_i) = \theta^T x_i$, the derivative becomes:

$$\nabla J(\theta) = \frac{2}{m} \sum_{i=1}^{m} (h_\theta(x_i) - y_i) x_i$$

2. **Solution:**

Given $h_\theta(x_i) = \frac{1}{1 + e^{-\theta^T x_i}}$, the derivative of the cross-entropy loss with respect to θ_j is:

$$\frac{\partial J}{\partial \theta_j} = -\frac{1}{m} \sum_{i=1}^{m} \left(y_i \frac{1}{h_\theta(x_i)} \frac{\partial h_\theta(x_i)}{\partial \theta_j} - (1 - y_i) \frac{1}{1 - h_\theta(x_i)} \frac{\partial (1 - h_\theta(x_i))}{\partial \theta_j} \right)$$

Resulting from:

$$\frac{\partial h_\theta(x_i)}{\partial \theta_j} = h_\theta(x_i)(1 - h_\theta(x_i)) x_{ij}$$

Thus:

$$\frac{\partial J}{\partial \theta_j} = \frac{1}{m} \sum_{i=1}^{m} (h_\theta(x_i) - y_i) x_{ij}$$

3. **Solution:**

For the regularized cost function:

$$J(\theta) = \frac{1}{m} \sum_{i=1}^{m} (h_\theta(x_i) - y_i)^2 + \lambda \sum_{j=1}^{n} \theta_j^2$$

The gradient:

$$\nabla J(\theta) = \frac{2}{m} \sum_{i=1}^{m} (h_\theta(x_i) - y_i) x_i + 2\lambda\theta$$

4. **Solution:**

The normal equation for linear regression is derived by setting the derivative of the cost function equal to zero:

$$X^T(X\theta - \mathbf{y}) = 0$$

Solving for θ:

$$\theta = (X^T X)^{-1} X^T \mathbf{y}$$

5. **Solution:**

The gradient descent update rule:

$$\theta := \theta - \alpha \nabla J(\theta)$$

Convergence is achieved if the learning rate α is chosen appropriately. The choice depends on the shape of the cost function. A too-large α could lead to divergence, while a too-small one results in slow convergence.

6. **Solution:**

The Hessian matrix $H = \nabla^2 J(\theta)$ indicates curvature:

$$\theta := \theta - H^{-1} \nabla J(\theta)$$

In Newton's method, accounting for curvature through the Hessian allows faster convergence to local minima, especially for quadratic-like surfaces. Positive definiteness ensures a unique minimum direction.

Chapter 25

Activation Functions and Their Derivatives

Practice Problems 1

1. Consider the sigmoid function $\texttt{sigmoid}(z) = \frac{1}{1+e^{-z}}$. Compute the second derivative $\frac{d^2}{dz^2}\texttt{sigmoid}(z)$.

2. Verify the identity for the derivative of the hyperbolic tangent function, $\frac{d}{dz}\texttt{tanh}(z) = 1 - \texttt{tanh}^2(z)$, by using the definition of $\texttt{tanh}(z)$.

3. Determine the Jacobian matrix J of the softmax function $\texttt{softmax}(\mathbf{z})$.

4. For ReLU activation, evaluate $\int \text{ReLU}(z)\,dz$.

5. Analyze the effect of α in the derivative of Leaky ReLU, $\frac{d}{dz}\text{LeakyReLU}(z)$, for optimization speed in gradient descent.

6. Prove that the softmax function outputs are invariant to constant shifts in the input vector, i.e., $\text{softmax}(\mathbf{z}+c\mathbf{1}) = \text{softmax}(\mathbf{z})$ for any constant c.

Answers 1

1. Compute the second derivative of the sigmoid function:

$$\text{sigmoid}(z) = \frac{1}{1+e^{-z}}$$

Solution: First derivative:

$$\frac{d}{dz}\text{sigmoid}(z) = \text{sigmoid}(z)(1-\text{sigmoid}(z))$$

Second derivative:

$$\frac{d^2}{dz^2}\text{sigmoid}(z) = \frac{d}{dz}\left[\text{sigmoid}(z)(1-\text{sigmoid}(z))\right]$$

$$= \text{sigmoid}'(z)(1-\text{sigmoid}(z)) - \text{sigmoid}(z)\text{sigmoid}'(z)$$

$$= \text{sigmoid}(z)(1-\text{sigmoid}(z))(1-2\text{sigmoid}(z))$$

2. Verify the hyperbolic tangent derivative identity:

$$\tanh(z) = \frac{e^z - e^{-z}}{e^z + e^{-z}}$$

Solution: Differentiate using quotient rule:

$$\frac{d}{dz}\tanh(z) = \frac{(e^z + e^{-z})(e^z + e^{-z}) - (e^z - e^{-z})(e^z - e^{-z})}{(e^z + e^{-z})^2}$$

Simplifying,

$$= \frac{4}{(e^z + e^{-z})^2}$$

$$= 1 - \left(\frac{e^z - e^{-z}}{e^z + e^{-z}}\right)^2 = 1 - \tanh^2(z)$$

3. Determine the Jacobian matrix of softmax:

$$\mathtt{softmax}(z_i) = \frac{e^{z_i}}{\sum_j e^{z_j}}$$

Solution: For $i = k$:

$$\frac{\partial}{\partial z_k}\mathtt{softmax}(z_i) = \mathtt{softmax}(z_i)(1 - \mathtt{softmax}(z_i))$$

For $i \neq k$:

$$\frac{\partial}{\partial z_k}\mathtt{softmax}(z_i) = -\mathtt{softmax}(z_i)\mathtt{softmax}(z_k)$$

Resulting matrix:

$$J_{ik} = \begin{cases} \mathtt{softmax}(z_i)(1 - \mathtt{softmax}(z_i)) & i = k \\ -\mathtt{softmax}(z_i)\mathtt{softmax}(z_k) & i \neq k \end{cases}$$

4. Evaluate the integral of ReLU function:

$$\mathtt{ReLU}(z) = \max(0, z)$$

Solution: For $z > 0$:

$$\int z\, dz = \frac{z^2}{2} + C$$

For $z \leq 0$, the integral is 0. Overall,

$$\int \mathtt{ReLU}(z)\, dz = \frac{z^2}{2}H(z) + C$$

Here $H(z)$ is the Heaviside step function.

5. Analyze effect of α on Leaky ReLU:

$$\frac{d}{dz}\mathtt{LeakyReLU}(z) = \begin{cases} 1 & z > 0 \\ \alpha & z \leq 0 \end{cases}$$

Solution: For $z > 0$, gradient $= 1$; no delay in convergence. For $z \leq 0$, gradient $= \alpha$; with smaller α slowing negative region adjustment. Larger α generally speeds up learning for negative activations in gradient descent.

6. Prove softmax invariance to constant shifts:

$$\text{softmax}(\mathbf{z} + c\mathbf{1}) = \frac{e^{z_i+c}}{\sum_j e^{z_j+c}} = \frac{e^{z_i}e^c}{e^c\sum_j e^{z_j}}$$

Solution: Simplifies to:

$$\frac{e^{z_i}}{\sum_j e^{z_j}}$$

Hence, $\text{softmax}(\mathbf{z} + c\mathbf{1}) = \text{softmax}(\mathbf{z})$, proving the shift invariance.

Practice Problems 2

1. Derive the expression for the derivative of the sigmoid activation function, $\sigma(z) = \frac{1}{1+e^{-z}}$, demonstrating each step clearly.

$$\text{Compute:} \quad \frac{d}{dz}\sigma(z)$$

2. Prove that the derivative of the hyperbolic tangent function, $\tanh(z) = \frac{e^z-e^{-z}}{e^z+e^{-z}}$, can be expressed as $1 - \tanh^2(z)$.

$$\text{Compute:} \quad \frac{d}{dz}\tanh(z)$$

3. Calculate the derivative of the Rectified Linear Unit (ReLU) function, $\text{ReLU}(z) = \max(0, z)$, and discuss its significance in neural network training.

$$\text{Compute:} \quad \frac{d}{dz}\text{ReLU}(z)$$

4. Derive the expression for the derivative of the softmax function with respect to its input, demonstrating why the result involves the Jacobian matrix.

$$\text{Given:} \quad \text{softmax}(z_i) = \frac{e^{z_i}}{\sum_j e^{z_j}}, \quad \text{compute:} \quad \frac{\partial}{\partial z_k}\text{softmax}(z_i)$$

5. Derive the gradient of the cost function for a neural network using the sigmoid activation function in the context of binary cross-entropy loss.

$$\text{Given:} \quad J(\theta) = -\frac{1}{m}\sum_{i=1}^{m}[y^{(i)}\log(\hat{y}^{(i)}) + (1 - y^{(i)})\log(1 - \hat{y}^{(i)})]$$

$$\text{where } \hat{y}^{(i)} = \sigma(z^{(i)}), \quad \text{compute:} \quad \frac{\partial J}{\partial \theta}$$

6. Analyze the stability offered by Leaky ReLU compared to standard ReLU in training deep neural networks, focusing on the impact of their respective derivatives.

$$\text{Discuss: why } \frac{d}{dz}\text{LeakyReLU}(z) \text{ with } \alpha > 0 \text{ addresses the zero-gradient issue in ReLU.}$$

Answers 2

1. Derive the expression for the derivative of the sigmoid activation function, $\sigma(z) = \frac{1}{1+e^{-z}}$.

Solution:

Using the chain rule,

$$\frac{d}{dz}\sigma(z) = \frac{d}{dz}\left(\frac{1}{1+e^{-z}}\right)$$

Let $u = 1 + e^{-z}$, then $\frac{d}{dz}u = -e^{-z}$.

$$\frac{d}{dz}\left(\frac{1}{u}\right) = -\frac{1}{u^2}\cdot\frac{d}{dz}u = -\frac{1}{(1+e^{-z})^2}\cdot(-e^{-z})$$

$$= \frac{e^{-z}}{(1+e^{-z})^2} = \sigma(z)(1-\sigma(z))$$

Therefore, the derivative is:

$$\frac{d}{dz}\sigma(z) = \sigma(z)(1-\sigma(z)).$$

2. Prove that the derivative of the hyperbolic tangent function, $\tanh(z) = \frac{e^z - e^{-z}}{e^z + e^{-z}}$, can be expressed as $1 - \tanh^2(z)$.

   Solution:

   Let $\tanh(z) = \frac{sinh(z)}{cosh(z)}$.

   Using the quotient rule:

   $$\frac{d}{dz}\tanh(z) = \frac{cosh(z)\cdot\frac{d}{dz}(sinh(z)) - sinh(z)\cdot\frac{d}{dz}(cosh(z))}{cosh^2(z)}$$

   Knowing $\frac{d}{dz}(sinh(z)) = cosh(z)$ and $\frac{d}{dz}(cosh(z)) = sinh(z)$,

   $$= \frac{cosh^2(z) - sinh^2(z)}{cosh^2(z)}$$

   Using the identity $cosh^2(z) - sinh^2(z) = 1$,

   $$= \frac{1}{cosh^2(z)}$$

   $$= 1 - \tanh^2(z)$$

   Therefore, $\frac{d}{dz}\tanh(z) = 1 - \tanh^2(z)$.

3. Calculate the derivative of the Rectified Linear Unit (ReLU) function, $\text{ReLU}(z) = \max(0, z)$.

   Solution:

   $$\text{ReLU}(z) = \begin{cases} z & \text{if } z > 0 \\ 0 & \text{otherwise} \end{cases}$$

   Derivative:

   $$\frac{d}{dz}\text{ReLU}(z) = \begin{cases} 1 & \text{if } z > 0 \\ 0 & \text{if } z \leq 0 \end{cases}$$

   Significance: The ReLU activation function addresses the vanishing gradient problem by maintaining a derivative of 1 for positive z, allowing weights to update effectively during backpropagation.

4. Derive the expression for the derivative of the softmax function.

   Solution:

   Given: $\text{softmax}(z_i) = \frac{e^{z_i}}{\sum_j e^{z_j}}$

   Using the chain rule and partial derivatives:

   $$\frac{\partial}{\partial z_k}\text{softmax}(z_i) = \begin{cases} \text{softmax}(z_i)(1 - \text{softmax}(z_i)) & \text{if } i = k \\ -\text{softmax}(z_i)\text{softmax}(z_k) & \text{otherwise} \end{cases}$$

   Resulting in a diagonal matrix (Jacobian) for cases where $i = k$ that simplifies computational gradients for backpropagation.

5. Derive the gradient of the cost function for a neural network using the sigmoid activation function.

   Solution:

   Cost function:

   $$J(\theta) = -\frac{1}{m}\sum_{i=1}^{m}[y^{(i)}\log(\hat{y}^{(i)}) + (1 - y^{(i)})\log(1 - \hat{y}^{(i)})]$$

   Substituting $\hat{y}^{(i)} = \sigma(z^{(i)})$:

   $$\frac{\partial J}{\partial \theta} = \frac{1}{m}\sum_{i=1}^{m}\left(\sigma(z^{(i)}) - y^{(i)}\right) \cdot x^{(i)}$$

   Employing the chain rule,

   $$\frac{\partial J}{\partial \theta} = (\sigma(z) - y) \cdot x$$

   It provides the necessary weight update during backpropagation by accounting for the error between prediction $\sigma(z)$ and actual output y.

6. Analyze the stability offered by Leaky ReLU compared to standard ReLU in neural networks.

   Solution:

   Leaky ReLU:

   $$\text{LeakyReLU}(z) = \begin{cases} z & \text{if } z > 0 \\ \alpha z & \text{otherwise} \end{cases}$$

   Derivative:

   $$\frac{d}{dz}\text{LeakyReLU}(z) = \begin{cases} 1 & \text{if } z > 0 \\ \alpha & \text{if } z \leq 0 \end{cases}$$

   Stability: Leaky ReLU addresses the dying ReLU problem by ensuring a small gradient (α) even for negative inputs, thus allowing non-zero gradients to flow through the network for all z and enabling better training in deep networks.

Practice Problems 3

1. Given the sigmoid function, verify that its derivative can be expressed as:

   $$\frac{d}{dz}\text{sigmoid}(z) = \text{sigmoid}(z)(1 - \text{sigmoid}(z))$$

2. Prove that the derivative of the hyperbolic tangent function is:

$$\frac{d}{dz}\tanh(z) = 1 - \tanh^2(z)$$

3. Calculate the derivative of the ReLU activation function and explain why it is not differentiable at $z = 0$:

$$\frac{d}{dz}\text{ReLU}(z) = \begin{cases} 1 & \text{if } z > 0 \\ 0 & \text{if } z < 0 \end{cases}$$

4. For the Leaky ReLU function, demonstrate the derivative formula and discuss its advantage over ReLU for $z < 0$:

$$\frac{d}{dz}\text{LeakyReLU}(z) = \begin{cases} 1 & \text{if } z > 0 \\ \alpha & \text{if } z \leq 0 \end{cases}$$

5. Derive the expression for the Jacobian matrix of the softmax function:

$$\frac{\partial}{\partial z_k}\text{softmax}(z_i) = \begin{cases} \text{softmax}(z_i)(1 - \text{softmax}(z_i)) & \text{if } i = k \\ -\text{softmax}(z_i)\text{softmax}(z_k) & \text{if } i \neq k \end{cases}$$

6. Discuss the implications of the vanishing gradient problem in sigmoid and tanh activation functions and why ReLU is often preferred in deep networks.

Answers 3

1. **Solution:** The sigmoid function is given by:

$$\text{sigmoid}(z) = \frac{1}{1 + e^{-z}}$$

Differentiate using the quotient rule:

$$\frac{d}{dz}\left(\frac{1}{1 + e^{-z}}\right) = \frac{0 \cdot (1 + e^{-z}) - (-e^{-z}) \cdot 1}{(1 + e^{-z})^2}$$

$$= \frac{e^{-z}}{(1 + e^{-z})^2}$$

$$= \frac{1}{1 + e^{-z}} \cdot \left(1 - \frac{1}{1 + e^{-z}}\right)$$

Therefore,

$$\frac{d}{dz}\text{sigmoid}(z) = \text{sigmoid}(z)(1 - \text{sigmoid}(z))$$

2. **Solution:** The hyperbolic tangent function is:

$$\tanh(z) = \frac{e^z - e^{-z}}{e^z + e^{-z}}$$

Differentiate using the quotient rule:

$$\frac{d}{dz}\left(\frac{e^z - e^{-z}}{e^z + e^{-z}}\right) = \frac{(e^z + e^{-z})^2 - (e^z - e^{-z})^2}{(e^z + e^{-z})^2}$$

Simplifying:

$$= \frac{4e^z e^{-z}}{(e^z + e^{-z})^2} = \frac{4}{(e^z + e^{-z})^2}$$

$$= 1 - \tanh^2(z)$$

3. **Solution:** The ReLU function is:

$$\text{ReLU}(z) = \max(0, z)$$

Derivative:

$$\frac{d}{dz}\text{ReLU}(z) = \begin{cases} 1 & \text{if } z > 0 \\ 0 & \text{if } z < 0 \end{cases}$$

At $z = 0$, it has a discontinuity in the derivative, making it non-differentiable at this point.

4. **Solution:** Leaky ReLU is defined as:

$$\text{LeakyReLU}(z) = \begin{cases} z & \text{if } z > 0 \\ \alpha z & \text{otherwise} \end{cases}$$

Derivative:

$$\frac{d}{dz}\text{LeakyReLU}(z) = \begin{cases} 1 & \text{if } z > 0 \\ \alpha & \text{if } z \leq 0 \end{cases}$$

Leaky ReLU addresses zero gradients by allowing a small gradient for $z < 0$.

5. **Solution:** For the softmax function:

$$\text{softmax}(z_i) = \frac{e^{z_i}}{\sum_j e^{z_j}}$$

The derivative involves:

$$\frac{\partial}{\partial z_k}\text{softmax}(z_i) = \text{softmax}(z_i)(\delta_{ik} - \text{softmax}(z_k))$$

Resulting in the specified Jacobian elements.

6. **Solution:** Sigmoid and tanh functions suffer from vanishing gradients for extreme input values, which slows learning. ReLU avoids this issue with linear behavior for $z > 0$, accelerating training, making it preferable in deeper architectures.

Chapter 26

Backpropagation Algorithm

Practice Problems 1

1. Explain the role of the Hadamard product in the chain rule application within the backpropagation algorithm:

$$\frac{\partial \mathcal{L}}{\partial \mathbf{z}^{(l)}} = \frac{\partial \mathcal{L}}{\partial \mathbf{a}^{(l)}} \circ \mathbf{f}'(\mathbf{z}^{(l)})$$

2. Derive the expression for updating the weights of a layer using the gradient descent rule in backpropagation:

$$\mathbf{W}^{(l)} := \mathbf{W}^{(l)} - \eta \frac{\partial \mathcal{L}}{\partial \mathbf{W}^{(l)}}$$

3. Demonstrate how the gradient of the loss with respect to the input to a particular layer is computed in backpropagation:

$$\frac{\partial \mathcal{L}}{\partial \mathbf{z}^{(l-1)}} = \left(\mathbf{W}^{(l)}\right)^{\top} \frac{\partial \mathcal{L}}{\partial \mathbf{z}^{(l)}} \circ \mathbf{f}'(\mathbf{z}^{(l-1)})$$

4. Discuss how initialization of gradients occurs at the output layer of the network during backpropagation:

$$\frac{\partial \mathcal{L}}{\partial \mathbf{z}^{(L)}} = \mathbf{a}^{(L)} - \mathbf{y}$$

5. Explain the significance of the learning rate η in the parameter update process of backpropagation:

$$\mathbf{W}^{(l)} := \mathbf{W}^{(l)} - \eta \frac{\partial \mathcal{L}}{\partial \mathbf{W}^{(l)}}$$

6. How does the backward pass of the backpropagation algorithm ensure that all parameters are being updated simultaneously across all layers?

Answers 1

1. **Solution:** The Hadamard product in the chain rule application within the backpropagation algorithm is used to ensure that the gradient computation is element-wise. When the derivative of the activation function $\mathbf{f}'(\mathbf{z}^{(l)})$ is involved, applying the Hadamard product ensures that each component of the gradient vector $\frac{\partial \mathcal{L}}{\partial \mathbf{a}^{(l)}}$ is multiplied by the corresponding derivative of the activation function. This ensures that the gradient flows back correctly and maintains the dimensions consistent with the parameter matrices. Therefore,

$$\frac{\partial \mathcal{L}}{\partial \mathbf{z}^{(l)}} = \frac{\partial \mathcal{L}}{\partial \mathbf{a}^{(l)}} \circ \mathbf{f}'(\mathbf{z}^{(l)}).$$

2. **Solution:** The gradient descent rule updates the weights to minimize the loss function. Starting with the gradient $\frac{\partial \mathcal{L}}{\partial \mathbf{W}^{(l)}}$ calculated during backpropagation, the update rule $\mathbf{W}^{(l)} := \mathbf{W}^{(l)} - \eta \frac{\partial \mathcal{L}}{\partial \mathbf{W}^{(l)}}$ applies the learning rate η to control the step size of the update. By subtracting $\eta \frac{\partial \mathcal{L}}{\partial \mathbf{W}^{(l)}}$ from the current weights, the rule moves towards a local minimum in the error landscape. This update is applied iteratively after each pass of the backpropagation. Therefore,

$$\mathbf{W}^{(l)} := \mathbf{W}^{(l)} - \eta \frac{\partial \mathcal{L}}{\partial \mathbf{W}^{(l)}}.$$

3. **Solution:** The gradient of the loss with respect to the input to a particular layer can be computed using the chain rule for vector derivatives. After calculating $\frac{\partial \mathcal{L}}{\partial \mathbf{z}^{(l)}}$, the derivative with respect to the input of the layer, $\frac{\partial \mathcal{L}}{\partial \mathbf{z}^{(l-1)}}$, is determined by matrix multiplication of the transposed weight matrix $\left(\mathbf{W}^{(l)}\right)^{\top}$ and the gradient at $\mathbf{z}^{(l)}$, element-wise multiplied (Hadamard product) by the derivative of the activation function $\mathbf{f}'(\mathbf{z}^{(l-1)})$. This backward propagation ensures that gradients allocated for each weight are propagated correctly to the input of the preceding layer. Therefore,

$$\frac{\partial \mathcal{L}}{\partial \mathbf{z}^{(l-1)}} = \left(\mathbf{W}^{(l)}\right)^{\top} \frac{\partial \mathcal{L}}{\partial \mathbf{z}^{(l)}} \circ \mathbf{f}'(\mathbf{z}^{(l-1)}).$$

4. **Solution:** During backpropagation, initialization of the gradient at the output layer is crucial for computing the subsequent gradients for earlier layers. Specifically, for a squared error loss function, the gradient of the loss with respect to the inputs to the output layer is calculated as $\frac{\partial \mathcal{L}}{\partial \mathbf{z}^{(L)}} = \mathbf{a}^{(L)} - \mathbf{y}$. This represents the difference between the predicted output $\mathbf{a}^{(L)}$ and the true target $\mathbf{y}$, which initiates the backward flow of gradients by defining the initial gradient value used for calculations in prior layers. Therefore,

$$\frac{\partial \mathcal{L}}{\partial \mathbf{z}^{(L)}} = \mathbf{a}^{(L)} - \mathbf{y}.$$

5. **Solution:** The learning rate η plays a critical role in the parameter update process by controlling the pace of these updates. A small η results in slow convergence, potentially leading to more accurate minima, while a large η may accelerate convergence but risk overshooting local minima, causing instability in training. Hence, choosing an appropriate η ensures efficient convergence towards minimal or satisfactory solutions in the model's error landscape. The weight update formula reflects how η regulates the gradient descent's scale:

$$\mathbf{W}^{(l)} := \mathbf{W}^{(l)} - \eta \frac{\partial \mathcal{L}}{\partial \mathbf{W}^{(l)}}.$$

6. **Solution:** In the backward pass of backpropagation, gradients are systematically propagated backwards from the output to the first layer, ensuring simultaneous updates for all parameters. This is achieved by using the chain rule to sequentially update each layer's parameters based on the gradient computed for the successive layer. Consequently, by initializing gradients at the output and meticulously applying derivatives back through each layer, backpropagation facilitates a synchronous update of weights and biases during gradient descent, optimizing the entire network holistically and efficiently for improved model performance.

Practice Problems 2

1. Explain the significance of the backpropagation algorithm in the training of neural networks, and derive the formula for updating weights $\mathbf{W}^{(l)}$ in gradient descent.

2. Given a neural network layer l with the activation function $\mathbf{f}(z) = \tanh(z)$, calculate the derivative of the activation with respect to its input z.

3. Describe the process of computing the gradient of the loss function with respect to the biases $\mathbf{b}^{(l)}$ for a layer l.

4. For a neural network using ReLU activation functions, calculate $\frac{\partial \mathcal{L}}{\partial \mathbf{z}^{(l)}}$ given $\mathbf{a}^{(l)} = \max(0, \mathbf{z}^{(l)})$ and $\mathcal{L} = \frac{1}{2}(\mathbf{a}^{(L)} - \mathbf{y})^2$.

5. How is the learning rate η chosen in the gradient descent algorithm, and how does it affect convergence when updating parameters?

6. Discuss the concept and significance of the Hadamard product (element-wise multiplication) in the backpropagation algorithm.

Answers 2

1. **Solution:** The backpropagation algorithm is vital because it allows the efficient computation of gradients for weights and biases in a neural network. By applying the chain rule of calculus backward through the network, it determines how much to adjust each parameter to minimize the loss function. For updating weights $\mathbf{W}^{(l)}$ in gradient descent:

$$\mathbf{W}^{(l)} := \mathbf{W}^{(l)} - \eta \frac{\partial \mathcal{L}}{\partial \mathbf{W}^{(l)}}$$

   This formula pertains to incrementally adjusting the weights in the direction that reduces the loss, scaled by the learning rate η.

2. **Solution:** The derivative of the hyperbolic tangent function is crucial in backpropagation:

$$\mathbf{f}(z) = \tanh(z) = \frac{e^z - e^{-z}}{e^z + e^{-z}}$$

$$\frac{\partial \mathbf{f}(z)}{\partial z} = 1 - \tanh^2(z)$$

   Deriving this result applies the derivative of the tanh function, leveraging its properties for efficient gradient computations.

3. **Solution:** The gradient with respect to biases $\mathbf{b}^{(l)}$ is straightforward due to the additive nature of bias:

$$\frac{\partial \mathcal{L}}{\partial \mathbf{b}^{(l)}} = \frac{\partial \mathcal{L}}{\partial \mathbf{z}^{(l)}}$$

   Since each element of $\mathbf{b}^{(l)}$ directly affects just one element in $\mathbf{z}^{(l)}$, the partial derivative transfers directly from the gradient of the layer above.

4. **Solution:** For ReLU, the derivative is stepwise:

$$\mathbf{a}^{(l)} = \max(0, \mathbf{z}^{(l)})$$

$$\frac{\partial \mathcal{L}}{\partial \mathbf{z}^{(l)}} = \begin{cases} \frac{\partial \mathcal{L}}{\partial \mathbf{a}^{(l)}} & \text{if } \mathbf{z}^{(l)} > 0 \\ 0 & \text{otherwise} \end{cases}$$

   This derives from the ReLU's nature: it's not differentiable at 0, so a subgradient is used during training.

5. **Solution:** Choosing η involves tuning: small values ensure convergence but may be slow, while large values can cause divergence or oscillation. Adaptive methods adjust η during training to stabilize convergence.

6. **Solution:** The Hadamard product $\circ$, or element-wise multiplication, is crucial in layer-wise calculation during backpropagation, allowing independent calculation of gradients across neuron outputs:

$$\frac{\partial \mathcal{L}}{\partial \mathbf{z}^{(l)}} = \frac{\partial \mathcal{L}}{\partial \mathbf{a}^{(l)}} \circ \mathbf{f}'(\mathbf{z}^{(l)})$$

   It calculates gradients efficiently when neural network operations maintain element-wise independence, key to parallel processing benefits.

Practice Problems 3

1. Consider a simple feedforward neural network with activation function derivatives. If the activation function at layer l is $\mathbf{f}(z) = \frac{1}{1+e^{-z}}$, find the derivative $\mathbf{f}'(z)$.

2. For a given neural network, if the loss function is given by $\mathcal{L} = \frac{1}{2}\|\mathbf{a}^{(L)} - \mathbf{y}\|^2$, derive the expression for $\frac{\partial \mathcal{L}}{\partial \mathbf{a}^{(L)}}$.

3. Given the gradient $\frac{\partial \mathcal{L}}{\partial \mathbf{z}^{(3)}}$ is known, show how to compute $\frac{\partial \mathcal{L}}{\partial \mathbf{a}^{(2)}}$ using the backpropagation method.

4. Explain how the Hadamard product is applied in the backpropagation algorithm when computing $\frac{\partial \mathcal{L}}{\partial \mathbf{z}^{(l)}}$ in a neural network.

5. Suppose a neural network uses the ReLU activation function, $\mathbf{f}(z) = \max(0, z)$. Derive the expression for $\mathbf{f}'(z)$ used in backpropagation.

6. Discuss the role of the learning rate η in the parameter update equations and its effect on the convergence of the neural network training.

Answers 3

1. **Solution:**

$$\mathbf{f}(z) = \frac{1}{1 + e^{-z}}$$

To find $\mathbf{f}'(z)$, use the derivative of the sigmoid function:

$$\mathbf{f}'(z) = \mathbf{f}(z)(1 - \mathbf{f}(z))$$

Therefore,

$$\mathbf{f}'(z) = \frac{1}{1 + e^{-z}}\left(1 - \frac{1}{1 + e^{-z}}\right).$$

2. **Solution:**

$$\mathcal{L} = \frac{1}{2}\|\mathbf{a}^{(L)} - \mathbf{y}\|^2$$

Differentiate with respect to $\mathbf{a}^{(L)}$:

$$\frac{\partial \mathcal{L}}{\partial \mathbf{a}^{(L)}} = \mathbf{a}^{(L)} - \mathbf{y}$$

Thus,

$$\frac{\partial \mathcal{L}}{\partial \mathbf{a}^{(L)}} = \mathbf{a}^{(L)} - \mathbf{y}.$$

3. **Solution:** Use the backpropagation method:

$$\frac{\partial \mathcal{L}}{\partial \mathbf{a}^{(2)}} = \left(\mathbf{W}^{(3)}\right)^{\top} \frac{\partial \mathcal{L}}{\partial \mathbf{z}^{(3)}}$$

Therefore,

$$\frac{\partial \mathcal{L}}{\partial \mathbf{a}^{(2)}} = \left(\mathbf{W}^{(3)}\right)^{\top} \frac{\partial \mathcal{L}}{\partial \mathbf{z}^{(3)}}$$

is calculated through the transposed weights.

4. **Solution:** The Hadamard product is critical in the layer-wise gradient computation:

$$\frac{\partial \mathcal{L}}{\partial \mathbf{z}^{(l)}} = \frac{\partial \mathcal{L}}{\partial \mathbf{a}^{(l)}} \circ \mathbf{f}'(\mathbf{z}^{(l)})$$

This operation ensures element-wise multiplication, aligning partial derivatives with specific activations.

5. **Solution:** Given ReLU, $\mathbf{f}(z) = \max(0, z)$:

$$\mathbf{f}'(z) = \begin{cases} 0 & \text{if } z \leq 0 \\ 1 & \text{if } z > 0 \end{cases}$$

ReLU's derivative is used to handle zeroed gradient flows.

6. **Solution:** Learning rate η determines the step size in parameter updates:

$$\mathbf{W}^{(l)} := \mathbf{W}^{(l)} - \eta \frac{\partial \mathcal{L}}{\partial \mathbf{W}^{(l)}}$$

$$\mathbf{b}^{(l)} := \mathbf{b}^{(l)} - \eta \frac{\partial \mathcal{L}}{\partial \mathbf{b}^{(l)}}$$

A small η might slow convergence, whereas a large η can cause overshooting or divergence.

Chapter 27

Optimization Challenges

Practice Problems 1

1. Consider the function $f(x,y) = x^3 - 3xy + y^3$. Determine if the point $(0,0)$ is a local minimum, local maximum, or a saddle point.

2. Using the Newton-Raphson method, solve for the root of the equation $f(x) = x^3 - 2x + 1 = 0$ starting from an initial guess $x_0 = 1$.

3. For the function $f(x,y) = x^2 + y^2$, calculate the Hessian and determine if the function is convex.

4. Given a function $f(x) = \frac{1}{3}x^3 - 4x + 1$, use calculus-based methods to find the critical points and determine their nature.

4. Consider the convex function $f(x) = x^2$. By definition of convexity, for any x_1, x_2 and $\lambda \in [0, 1]$, we have:

$$f(\lambda x_1 + (1-\lambda)x_2) \leq \lambda f(x_1) + (1-\lambda)f(x_2)$$

Substituting $f(x) = x^2$:

$$(\lambda x_1 + (1-\lambda)x_2)^2 \leq \lambda x_1^2 + (1-\lambda)x_2^2$$

This inequality holds, confirming $f(x) = x^2$ is convex.

5. We solve the constrained optimization problem using the method of Lagrange multipliers. The Lagrangian is:

$$\mathcal{L}(x, y, \lambda) = xy - \lambda(x^2 + y^2 - 1)$$

Taking partial derivatives and setting them to zero:

$$\frac{\partial \mathcal{L}}{\partial x} = y - 2\lambda x = 0$$

$$\frac{\partial \mathcal{L}}{\partial y} = x - 2\lambda y = 0$$

$$\frac{\partial \mathcal{L}}{\partial \lambda} = -(x^2 + y^2 - 1) = 0$$

From the first two equations, we get $y = 2\lambda x$ and $x = 2\lambda y$. Solving these, we find $x = \pm y$. Substituting into the constraint $x^2 + y^2 = 1$, we get $x = \pm \frac{1}{\sqrt{2}}$ and $y = \pm \frac{1}{\sqrt{2}}$. The extrema are at $\left(\frac{1}{\sqrt{2}}, \frac{1}{\sqrt{2}}\right)$ and $\left(-\frac{1}{\sqrt{2}}, -\frac{1}{\sqrt{2}}\right)$.

6. Stochastic gradient descent (SGD) updates the model parameters using a randomly selected subset of the data at each step. This randomness introduces noise into the optimization process, which can help the algorithm escape local minima in non-convex optimization problems. In scenarios with large datasets or complex loss surfaces, such as training deep neural networks, SGD is advantageous because it allows for faster convergence and better generalization by avoiding overfitting to local minima.

Questions 2

1. Given the function $f(x, y) = x^2 + y^2 - 4x - 6y + 13$, find the minimum value using gradient descent. Start from the point $(0, 0)$ with a learning rate of 0.1 and perform three iterations.

2. Consider the function $f(x) = x^4 - 3x^3 + 2$. Find the critical points and classify them using the second derivative test.

260

4. Differentiating $f(x) = \frac{1}{3}x^3 - 4x + 1$, we find the first derivative:

$$f'(x) = x^2 - 4$$

Setting $f'(x) = 0$ gives $x^2 = 4$, so $x = \pm 2$.

The second derivative $f''(x) = 2x$ evaluated at these points:

$f''(2) = 4$ (local minimum), $f''(-2) = -4$ (local maximum).

5. Using Lagrange multipliers, given $L(x, y, \lambda) = xy + \lambda(x^2 + y^2 - 1)$, finding the extrema:

$$\frac{\partial L}{\partial x} = y + 2\lambda x = 0$$

$$\frac{\partial L}{\partial y} = x + 2\lambda y = 0$$

$$\frac{\partial L}{\partial \lambda} = x^2 + y^2 - 1 = 0$$

Assume $\lambda = 0$ results in $x = y = 0$, which does not satisfy the constraint.

Solving the other equations gives $x = \pm\frac{\sqrt{2}}{2}$, $y = \pm\frac{\sqrt{2}}{2}$.

Therefore, the extrema points are $\left(\frac{\sqrt{2}}{2}, \frac{\sqrt{2}}{2}\right)$ and $\left(-\frac{\sqrt{2}}{2}, -\frac{\sqrt{2}}{2}\right)$.

6. Stochastic gradient descent (SGD) introduces randomness by updating parameters using a randomly selected subset (mini-batch) of data rather than the entire dataset, which helps escape local minima due to the noisy gradients.

   A scenario where SGD is advantageous: training deep neural networks, as the noise might help traverse complex cost surfaces with numerous local minima, potentially finding better regions in the parameter space.

Practice Problems 2

1. Explain how the gradient descent algorithm might fail in finding a global minimum due to the presence of local minima. Provide a conceptual example involving a non-convex function.

2. Given a function $f(x, y) = x^2 + y^2 - xy - 2x + y$, determine whether the point $(1, 1)$ is a local minimum, maximum, or a saddle point.

3. Describe how the Newton-Raphson method uses the Hessian matrix and derive the update formula for a univariate function.

4. Discuss the role of stochastic gradient descent (SGD) in overcoming the challenge posed by saddle points. Illustrate with a scenario.

5. Calculate the Hessian matrix for the function $f(x, y) = 3x^2 + y^3 - 5xy$ and determine the nature of the critical point at $(0, 0)$.

6. Explain how adaptive learning rates help in addressing optimization challenges and provide an example calculation for a simple logistic regression problem.

Answers 2

1. **Solution:** In the gradient descent algorithm, each update step involves moving in the direction of the steepest descent based on the gradient calculation. Due to the presence of local minima in non-convex

functions, the algorithm may prematurely converge to a local minimum instead of the global minimum. An example is the function $f(x) = x^4 - x^2 + x$, exhibiting multiple local minima.

In such cases, the local gradient may be zero, indicating a minimum without reaching the global minimum. Visualizing the function graph can help in understanding these local minima traps.

2. **Solution:** To determine the nature of the point $(1,1)$, calculate the first and second derivatives:

$$f(x,y) = x^2 + y^2 - xy - 2x + y$$

$$\frac{\partial f}{\partial x} = 2x - y - 2, \quad \frac{\partial f}{\partial y} = 2y - x + 1$$

At $(1,1)$, both partials become zero. Next, compute the Hessian:

$$\mathbf{H} = \begin{bmatrix} \frac{\partial^2 f}{\partial x^2} & \frac{\partial^2 f}{\partial x \partial y} \\ \frac{\partial^2 f}{\partial y \partial x} & \frac{\partial^2 f}{\partial y^2} \end{bmatrix} = \begin{bmatrix} 2 & -1 \\ -1 & 2 \end{bmatrix}$$

The determinant of the Hessian $\det(\mathbf{H}) = 3$ and is positive, with positive leading diagonal elements, indicating a local minimum at $(1,1)$.

3. **Solution:** The Newton-Raphson method uses the Hessian $\mathbf{H}$ to adjust the gradient descent step for improved convergence:

$$\mathbf{x}_{k+1} = \mathbf{x}_k - \mathbf{H}^{-1}(\mathbf{x}_k)\nabla f(\mathbf{x}_k)$$

For a univariate function $f(x)$, if $f''(x)$ is the second derivative, the update formula simplifies to:

$$x_{k+1} = x_k - \frac{f'(x_k)}{f''(x_k)}$$

This formula uses the curvature information (via the second derivative) to more accurately find the root/minimum point, assuming convergence conditions are met.

4. **Solution:** Stochastic Gradient Descent (SGD) introduces randomness by selecting random subsets (mini-batches) of the data to calculate the gradient, rather than the full dataset. This can help escape saddle points due to variability in gradient updates. For example, consider optimizing $f(x) = x^3 - 3x$. At $x = 0$, the gradient is zero, indicating a saddle point. Using SGD allows variability in steps due to random sampling, aiding in navigating away from the saddle point.

5. **Solution:** For the function $f(x,y) = 3x^2 + y^3 - 5xy$, the Hessian is:

$$\mathbf{H} = \begin{bmatrix} \frac{\partial^2 f}{\partial x^2} & \frac{\partial^2 f}{\partial x \partial y} \\ \frac{\partial^2 f}{\partial y \partial x} & \frac{\partial^2 f}{\partial y^2} \end{bmatrix} = \begin{bmatrix} 6 & -5 \\ -5 & 6y \end{bmatrix}$$

At $(0,0)$, the Hessian becomes:

$$\mathbf{H} = \begin{bmatrix} 6 & -5 \\ -5 & 0 \end{bmatrix}$$

The determinant is $\det(\mathbf{H}) = -30$, indicating both negative and positive eigenvalues, thus confirming a saddle point at $(0,0)$.

6. **Solution:** Adaptive learning rates adjust the learning rate η dynamically, benefiting optimization by adjusting to the function's curvature or plateau areas. For instance, in logistic regression,

$$\text{Model: } \sigma(\mathbf{W}\mathbf{x} + b), \quad \text{Loss: } L(\mathbf{W})$$

Suppose $\nabla L(\mathbf{W})$ is the gradient. An adaptive approach, like Adam, uses:

$$\eta_t = \frac{\alpha}{\sqrt{\hat{v}_t} + \epsilon}$$

where $\hat{v}_t$ represents a moving average of past squared gradients, refining learning stability, flaring the rate when gradients stabilize, and adjusting when needed during convergence challenges.

Practice Problems 3

1. Prove that the Newton-Raphson update rule for a function $f : \mathbb{R} \to \mathbb{R}$ will converge to a local minimum only if the initial guess is sufficiently close to the local minimum.

2. Consider the function $f(x, y) = x^2 + y^4 - 4xy$. Determine the nature of the critical points of f using the Hessian matrix.

3. Demonstrate that the function $g(x) = \frac{x^3}{3} - x$ has a saddle point.

4. Find the function $h(x, y) = x^3 - 3xy + y^3$ has any local minima, maxima, or saddle points.

5. Use the gradient descent method numerically to approximate the minimum of $f(x) = x^4 - 4x^2 + 4$.

6. Given a function $f(x) = e^{-x^2}$, explain why the function does not have any local minima or maxima.

Answers 3

1. **Solution:** The Newton-Raphson method is given by:

$$x_{k+1} = x_k - \frac{f'(x_k)}{f''(x_k)}$$

For convergence to a local minimum, $f''(x^*) > 0$ at the local minimum x^*. If the initial guess x_0 is not close to x^*, the quadratic approximation $f(x) \approx f(x_k) + f'(x_k)(x - x_k) + \frac{f''(x_k)}{2}(x - x_k)^2$ may not be valid, leading to divergence or convergence to a different point. The method relies on $f''(x)$ being approximately constant near the minimum, which requires x_0 to be close enough.

2. **Solution:** Find the partial derivatives and set them to zero:

$$\frac{\partial f}{\partial x} = 2x - 4y, \quad \frac{\partial f}{\partial y} = 4y^3 - 4x$$

Setting these to zero gives critical points. Solve:

$$2x - 4y = 0 \quad \Rightarrow \quad x = 2y$$

$$4y^3 - 4x = 0 \quad \Rightarrow \quad 4y^3 - 4(2y) = 0 \quad \Rightarrow \quad y(y^2 - 2) = 0$$

Thus, critical points are $(0,0), (2\sqrt{2}, \sqrt{2}), (-2\sqrt{2}, -\sqrt{2})$. Evaluate Hessian $\mathbf{H}$:

$$\mathbf{H} = \begin{bmatrix} 2 & -4 \\ -4 & 12y^2 \end{bmatrix}$$

Calculate determinant for each critical point to classify.

3. **Solution:** Take derivatives:

$$g'(x) = x^2 - 1$$
$$g''(x) = 2x$$

The critical point is $x = \pm 1$ (found by solving $g'(x) = 0$). Evaluate second derivative: $g''(x) = 0$ at $x = 0$, indicating a possible saddle point. Solve for values to confirm saddle characteristics by checking concavity changes.

4. **Solution:** Compute gradient ∇h:

$$\frac{\partial h}{\partial x} = 3x^2 - 3y, \quad \frac{\partial h}{\partial y} = 3y^2 - 3x$$

Set the gradient to zero and solve:

$$3x^2 = 3y, \quad 3y^2 = 3x$$

Transform the system to find $x = y^2$ and solve. Evaluate and find critical points, classify using the Hessian.

264

5. **Solution:** Use initial guess x_0, tolerance ϵ, etc. Update rule for gradient descent:

$$x_{n+1} = x_n - \eta \nabla f(x_n)$$

$$\nabla f(x) = 4x^3 - 8x$$

Numerically iterate to find approximate values and compare against known values.

6. **Solution:** Calculate derivative:

$$f'(x) = -2xe^{-x^2}$$

Set $f'(x) = 0$ only at $x = 0$. Evaluate second derivative:

$$f''(x) = (4x^2 - 2)e^{-x^2}$$

Thus, no change in sign indicates neither maxima nor minima for function overall. Thus, it fits a Gaussian distribution, always concave.

Chapter 28

Hyperparameter Tuning

Practice Problems 1

1. Explain the importance of selecting an appropriate learning rate η using the convergence properties of gradient descent. What potential issues arise from η being too large or too small?

2. Consider a regularized cost function given by $J(\theta) + \lambda R(\theta)$. Analyze the impact of varying the regularization parameter λ on the bias-variance tradeoff of the model.

3. Derive the update rule for parameters using mini-batch gradient descent and specify how batch size m influences convergence and computational efficiency.

4. Calculate the sensitivity of the cost function $C(\theta, \eta)$ to changes in the learning rate η using calculus. Express it in terms of partial derivatives.

5. Conduct a scenario analysis where the Hessian matrix $\mathbf{H}$ of a cost function is used to explain the effect of hyperparameter tuning on convergence rates.

6. Illustrate how grid search and random search methods can be practically implemented for hyperparameter tuning in machine learning, and discuss their relative advantages and disadvantages.

Answers 1

1. **Solution:**

    A proper learning rate η ensures that gradient descent updates converge efficiently to a minimum. If η is too large, updates may overshoot the minimum leading to divergence or oscillation. If η is too small, convergence is excessively slow, risking getting stuck in local minima. Thus, η should be chosen balancing speed and stability.

2. **Solution:**

    Increasing λ affects the bias-variance tradeoff: it generally increases bias and decreases variance as the model becomes simpler, potentially leading to underfitting. Conversely, lowering λ reduces bias and increases variance, which could result in overfitting. Thus, λ should be tuned to find an optimal balance for generalization.

3. **Solution:**

    In mini-batch gradient descent, the update rule for parameters θ is

$$\theta := \theta - \eta \frac{1}{m} \sum_{i=1}^{m} \nabla J(\theta; x^{(i)}).$$

Here, batch size m influences the frequency and size of updates. Smaller m leads to noisy updates which might help in escaping local minima but also might miss the global minimum. Larger m offers smooth convergence with higher computational cost per iteration.

4. **Solution:**

The sensitivity of the cost function $C(\theta, \eta)$ to η is evaluated using the partial derivative:

$$\frac{\partial C}{\partial \eta}.$$

This sensitivity analysis helps understand how changes in η affect C. If $\frac{\partial C}{\partial \eta}$ is large, small variations in η can cause significant changes in cost, necessitating careful tuning.

5. **Solution:**

The Hessian $\mathbf{H}$ describes the local curvature of the cost function. If eigenvalues of $\mathbf{H}$ are large, the function has steep slopes, suggesting small η for stable convergence. Small eigenvalues indicate flat regions where larger η might speed up convergence without risking divergence.

6. **Solution:**

Grid search explores hyperparameter space systematically but can be computationally expensive as dimensionality grows. It guarantees a thorough evaluation across defined boundaries. *Random search* examines hyperparameters randomly, requiring fewer evaluations to achieve comparable performance, particularly useful in high-dimensional spaces. Each has its use-case depending on computational resources and parameter landscape.

Practice Problems 2

1. Given the cost function $C(\theta, \eta) = \theta^2 + 3\eta^2 + 2\theta\eta$, find the sensitivity of the cost function with respect to the learning rate η.

2. For a regularization term $R(\theta) = \lambda \left(\frac{1}{2}\theta^2 \right)$, find the expression for the gradient with respect to θ and analyze its impact when λ is varied.

3. Analyze the relationship between the batch size m and convergence rate by considering the update rule $\theta := \theta - \eta \frac{1}{m} \sum_{i=1}^{m} \nabla J(\theta; x^{(i)})$. Discuss how calculus helps to understand the impact of varying m.

4. Determine the role of the second derivative (Hessian) of a function in the convergence of hyperparameter tuning by examining $H = \begin{bmatrix} 6 & 2 \\ 2 & 6 \end{bmatrix}$. Discuss how the choice of learning rate η affects convergence.

5. Use calculus to derive the optimum learning rate η^* for gradient descent by considering the simple quadratic cost function $J(\theta) = \frac{1}{2}\theta^2$. Assume $J'(\theta) = \theta$.

6. Explain the importance of sensitivity analysis in hyperparameter tuning using the sample cost function $C(\theta, \eta, \lambda) = \theta^2 + \eta^2 + \lambda^2$. Find $\frac{\partial C}{\partial \lambda}$ and interpret its implications.

Answers 2

1. For $C(\theta, \eta) = \theta^2 + 3\eta^2 + 2\theta\eta$, we differentiate with respect to η:

$$\frac{\partial C}{\partial \eta} = \frac{\partial}{\partial \eta}(\theta^2 + 3\eta^2 + 2\theta\eta)$$

$$= 0 + 6\eta + 2\theta.$$

Therefore, the sensitivity of C to η is $6\eta + 2\theta$.

2. For $R(\theta) = \lambda\left(\frac{1}{2}\theta^2\right)$, the gradient with respect to θ is:

$$\frac{dR}{d\theta} = \frac{d}{d\theta}\left(\lambda \cdot \frac{1}{2}\theta^2\right)$$

$$= \lambda \cdot \theta.$$

When λ increases, the regularization effect is stronger, leading to more penalized θ values, affecting model complexity and overfitting.

3. For the update rule $\theta := \theta - \eta\frac{1}{m}\sum_{i=1}^{m}\nabla J(\theta; x^{(i)})$, increasing the batch size m:

$$\frac{1}{m}\sum_{i=1}^{m}\nabla J(\theta; x^{(i)})$$

reduces variance in gradient estimation, leading to more stable convergence but potentially slower per epoch. Calculus helps quantify changes to convergence through derivative analysis.

4. The Hessian matrix $H = \begin{bmatrix} 6 & 2 \\ 2 & 6 \end{bmatrix}$, indicates curvature:

$$\lambda_1, \lambda_2 = 6 \pm 2.$$

With eigenvalues 8 and 4, the choice of η should balance these values to ensure stable convergence, avoiding overshooting or under-correction.

5. For $J(\theta) = \frac{1}{2}\theta^2$, the gradient is $J'(\theta) = \theta$. The optimal η^* minimizes

$$J(\theta - \eta^* J'(\theta)) \approx J(\theta) - \eta^*\theta \cdot \theta + \frac{1}{2}\eta^{*2}\theta^2$$

Simplifying,

$$\frac{d}{d\eta^*}\left(-\eta^*\theta^2 + \frac{1}{2}\eta^{*2}\theta^2\right) = 0$$

gives $\eta^* = \frac{1}{\theta^2}$.

6. For $C(\theta, \eta, \lambda) = \theta^2 + \eta^2 + \lambda^2$, differentiation yields:

$$\frac{\partial C}{\partial \lambda} = 2\lambda.$$

Large $\frac{\partial C}{\partial \lambda}$ suggests that small changes in λ significantly affect C, emphasizing careful selection in tuning for minimal cost.

Practice Problems 3

1. Consider a cost function $C(\theta, \eta) = \theta^2 + e^{\eta\theta} + \ln(\theta + 1)$. Compute $\frac{\partial C}{\partial \eta}$ and interpret its significance in terms of sensitivity analysis.

2. Given the hyperparameter-tuned cost function $J(\theta, \lambda) = \frac{1}{2}(\theta - 3)^2 + \lambda\theta$, determine $\frac{\partial J}{\partial \lambda}$ and explain its implications for regularization.

3. If the cost function $F(\theta, \eta, m) = \frac{1}{m} \sum_{i=1}^{m} \left(\theta x^{(i)} - y^{(i)}\right)^2$ is affected by batch size m, find $\frac{\partial F}{\partial m}$ and discuss how changes in m could influence optimization stability.

4. Analyze the convergence condition of the cost function $G(\theta) = 5\theta^2 + \eta\theta^4$ by examining the second derivative. Discuss the implications of the learning rate η on the convergence speed.

5. For the cost function influenced by regularization $H(\theta, \lambda) = \theta^2 + \lambda(\theta - 1)^4$, find the critical points with respect to θ and analyze the role of λ in determining these points.

6. Suppose the gradient of a neural network is given by $\nabla_\theta J(\theta) = \begin{bmatrix} \theta_1^2 \\ 2\theta_2 \end{bmatrix}$. Compute the Hessian matrix and explain how it can assist in selecting hyperparameters for optimization.

Answers 3

1. Compute $\frac{\partial C}{\partial \eta}$ for $C(\theta, \eta) = \theta^2 + e^{\eta\theta} + \ln(\theta + 1)$.
   Solution:
   $$\frac{\partial C}{\partial \eta} = \frac{\partial}{\partial \eta}\left(\theta^2 + e^{\eta\theta} + \ln(\theta + 1)\right)$$
   $$= 0 + \theta e^{\eta\theta} + 0 = \theta e^{\eta\theta}$$

   The derivative $\theta e^{\eta\theta}$ shows how changes in the hyperparameter η affect the cost C. A large derivative implies high sensitivity to η.

2. Determine $\frac{\partial J}{\partial \lambda}$ for $J(\theta, \lambda) = \frac{1}{2}(\theta - 3)^2 + \lambda\theta$.
   Solution:
   $$\frac{\partial J}{\partial \lambda} = \frac{\partial}{\partial \lambda}\left(\frac{1}{2}(\theta - 3)^2 + \lambda\theta\right)$$
   $$= 0 + \theta = \theta$$

   This solution implies that the cost function J is linearly increasing with λ, showing a direct impact of regularization on the model output.

3. Find $\frac{\partial F}{\partial m}$ for $F(\theta, \eta, m) = \frac{1}{m}\sum_{i=1}^{m}\left(\theta x^{(i)} - y^{(i)}\right)^2$.
   Solution:
   $$\frac{\partial F}{\partial m} = -\frac{1}{m^2}\sum_{i=1}^{m}\left(\theta x^{(i)} - y^{(i)}\right)^2$$

   The result suggests that increasing m can decrease F, indicating more stable convergence with larger batch sizes.

4. Examine the convergence condition of $G(\theta) = 5\theta^2 + \eta\theta^4$.
   Solution:
   $$G''(\theta) = \frac{d^2}{d\theta^2}(5\theta^2 + \eta\theta^4) = 10 + 12\eta\theta^2$$

   The second derivative shows positivity, ensuring local minima. A smaller η is recommended for faster convergence.

5. Find critical points of $H(\theta, \lambda) = \theta^2 + \lambda(\theta - 1)^4$ with respect to θ.
   Solution:
   $$\frac{\partial H}{\partial \theta} = 2\theta + 4\lambda(\theta - 1)^3$$

   Setting $\frac{\partial H}{\partial \theta} = 0$,
   $$2\theta + 4\lambda(\theta - 1)^3 = 0 \Rightarrow \theta(2 + 4\lambda(\theta - 1)) \equiv 0$$

   Solving gives critical points $\theta = 0$ or $\theta = 1$, affected by λ.

6. Compute the Hessian for $\nabla_\theta J(\theta) = \begin{bmatrix} \theta_1^2 \\ 2\theta_2 \end{bmatrix}$.

   Solution:

$$H(\theta) = \begin{bmatrix} \frac{\partial^2}{\partial \theta_1^2}(\theta_1^2) & \frac{\partial^2}{\partial \theta_1 \partial \theta_2}(\theta_1^2) \\ \frac{\partial^2}{\partial \theta_2 \partial \theta_1}(2\theta_2) & \frac{\partial^2}{\partial \theta_2^2}(2\theta_2) \end{bmatrix}$$

$$= \begin{bmatrix} 2\theta_1 & 0 \\ 0 & 2 \end{bmatrix}$$

The Hessian matrix helps us understand the curvature, offering insights into stable hyperparameter selection for each direction.

Chapter 29

Momentum and Adaptive Learning Rates

Practice Problems 1

1. Show that the momentum update rule integrates the history of gradients and explain its effect on the parameter space navigation.

$$\mathbf{v} := \beta \mathbf{v} - \eta \nabla J(\theta)$$

$$\theta := \theta + \mathbf{v}$$

2. Compare the convergence properties of standard gradient descent and momentum-based gradient descent by considering an ill-conditioned quadratic bowl.

3. Derive the update rule for the Nesterov Accelerated Gradient (NAG) and discuss its advantages over classical momentum.

$$\mathbf{v} := \beta \mathbf{v} - \eta \nabla J(\theta + \beta \mathbf{v})$$

4. Explain how adaptive learning rates help in optimizing deep networks with diverse parameters, using Adagrad as an example.

$$\theta_i := \theta_i - \frac{\eta}{\sqrt{G_{ii} + \epsilon}} \nabla_{\theta_i} J(\theta)$$

5. Analyze the trade-offs involved in using RMSprop over Adagrad by examining their respective update rules.

6. Derive the bias correction terms in Adam and explain their significance in the algorithm's convergence behavior.

$$\hat{m}_t := \frac{m_t}{1 - \beta_1^t}, \quad \hat{v}_t := \frac{v_t}{1 - \beta_2^t}$$

Answers 1

1. **Solution:** The momentum update rule leverages past gradients, leading to a velocity vector accumulation.

$$\mathbf{v}_t = \beta \mathbf{v}_{t-1} - \eta \nabla J(\theta_{t-1})$$

Given its recursive nature, $\mathbf{v}_t$ effectively sums up past gradients weighted by β^k.

$$\mathbf{v}_t = -\eta \sum_{k=0}^{t-1} \beta^k \nabla J(\theta_{t-1-k})$$

Thus, it helps in accelerating convergence through a smoother, velocity-aided path effectively reducing oscillations on narrow, elongated regions of the parameter space.

2. **Solution:** In a quadratic bowl, standard gradient descent can be hindered by oscillations, particularly when ill-conditioned:

$$\theta := \theta - \eta \nabla J(\theta)$$

Convergence speed varies greatly along different dimensions due to varied scaling, making the trajectory zigzag. Momentum-based descent:

$$\theta := \theta + \mathbf{v}, \quad \mathbf{v} := \beta \mathbf{v} - \eta \nabla J(\theta)$$

smooths the oscillations due to its accumulated velocity, resulting in faster convergence and better handling of scale differences across dimensions.

3. **Solution:** NAG improves upon momentum by preemptively using future position information for its gradient evaluation.

$$\theta'_t := \theta_{t-1} + \beta \mathbf{v}_{t-1}$$
$$\mathbf{v}_t := \beta \mathbf{v}_{t-1} - \eta \nabla J(\theta'_t)$$

By computing the gradient $\nabla J(\theta'_t)$, NAG anticipates more precisely, especially beneficial for complex landscapes, enhancing convergence speed.

4. **Solution:** Adagrad's adaptation of learning rates affects parameter updates:

$$\theta_i := \theta_i - \frac{\eta}{\sqrt{G_{ii} + \epsilon}} \nabla_{\theta_i} J(\theta)$$

where G_{ii} accumulates squared gradients. This scaling ensures more significant steps for infrequently updated parameters, aiding deep networks that have varying useful features across layers.

5. **Solution:** RMSprop improves upon Adagrad by addressing diminishing rate issues:

$$\mathbf{E}[g^2]_t := \gamma \mathbf{E}[g^2]_{t-1} + (1 - \gamma)(\nabla_{\theta_i} J(\theta))^2$$
$$\theta_i := \theta_i - \frac{\eta}{\sqrt{\mathbf{E}[g^2]_t + \epsilon}} \nabla_{\theta_i} J(\theta)$$

Contrast: RMSprop maintains a decaying-average which retains adaptive learning advantage while controlling the rate reduction over time.

6. **Solution:** Adam's bias correction ensures unbiased first and second moment estimates:

$$m_t = \beta_1 m_{t-1} + (1 - \beta_1) \nabla J(\theta_t)$$
$$\hat{m}_t = \frac{m_t}{1 - \beta_1^t}$$

Similar $\hat{v}_t$ for the second moment. Significance: This correction counters initialization bias, crucial during initial convergence phases, granting Adam robust and steady learning behaviors.

Practice Problems 2

1. Explain the concept of momentum in optimization and provide a mathematical expression for how momentum modifies the standard gradient descent update.

2. Derive the Nesterov Accelerated Gradient (NAG) update rule starting from the standard momentum update rule.

3. Compare and contrast Adagrad and RMSprop. In your explanation, include how each method adjusts the parameter updates based on historical gradients.

4. Demonstrate the parameter update rule for the Adam optimizer and explain how bias-corrected estimates are used in Adam's adjustments.

5. Calculate the velocity term $\mathbf{v}$ after two iterations using momentum with $\beta = 0.9$, given initial velocity $\mathbf{v}_0 = 0$, learning rate $\eta = 0.01$, and gradients $\nabla J(\theta)$ in the first and second iteration are -3 and -2, respectively.

6. For the following sequence of gradients $[2, 4, 6]$, calculate the adjusted learning rates for each iteration using Adagrad with an initial learning rate $\eta = 0.01$ and $\epsilon = 10^{-8}$.

Answers 2

1. **Explain the concept of momentum in optimization and provide a mathematical expression for how momentum modifies the standard gradient descent update.**

   Solution: Momentum in optimization is used to accelerate gradient descent by accumulating an exponentially decaying moving average of past gradients and continuing in their direction. It reduces oscillations and enhances convergence on surfaces that are poorly scaled. The momentum update rule is:

   $$\mathbf{v} := \beta\mathbf{v} - \eta\nabla J(\theta),$$
   $$\theta := \theta + \mathbf{v},$$

   where β is the momentum coefficient, η is the learning rate, and $\nabla J(\theta)$ is the gradient of the cost function with respect to parameters θ.

2. **Derive the Nesterov Accelerated Gradient (NAG) update rule starting from the standard momentum update rule.**

   Solution: In standard momentum, the gradient is computed at the current parameters. Nesterov Accelerated Gradient (NAG) computes the gradient at the projected position, thus:

   $$\mathbf{v} = \beta\mathbf{v} - \eta\nabla J(\theta + \beta\mathbf{v}),$$
   $$\theta = \theta + \mathbf{v}.$$

   This allows a look-ahead strategy which can result in faster convergence by effectively using information about the "future" location.

3. **Compare and contrast Adagrad and RMSprop. In your explanation, include how each method adjusts the parameter updates based on historical gradients.**

   Solution: Adagrad:

   $$\theta_i := \theta_i - \frac{\eta}{\sqrt{G_{ii} + \epsilon}}\nabla_{\theta_i} J(\theta),$$

   where G is the sum of squares of past gradients. It scales the learning rate inversely with the accumulation of the past gradient magnitudes, equalizing learning rates across all parameters.

   RMSprop:

   $$\mathbf{E}[g^2]_t = \gamma\mathbf{E}[g^2]_{t-1} + (1 - \gamma)(\nabla_{\theta_i} J(\theta))^2,$$
   $$\theta_i = \theta_i - \frac{\eta}{\sqrt{\mathbf{E}[g^2]_t + \epsilon}}\nabla_{\theta_i} J(\theta).$$

   It uses an exponentially decaying average of gradients' squared magnitudes, preventing the learning rate from vanishing and stabilizing updates. RMSprop adjusts Adagrad by introducing a decay factor to control the past gradients' accumulation, which is particularly beneficial for handling non-convex problems.

4. **Demonstrate the parameter update rule for the Adam optimizer and explain how bias-corrected estimates are used in Adam's adjustments.**

   Solution: Adam combines momentum and RMSprop by computing moment estimates:

   $$m_t = \beta_1 m_{t-1} + (1 - \beta_1)\nabla J(\theta),$$
   $$v_t = \beta_2 v_{t-1} + (1 - \beta_2)(\nabla J(\theta))^2,$$
   $$\hat{m}_t = \frac{m_t}{1 - \beta_1^t},$$
   $$\hat{v}_t = \frac{v_t}{1 - \beta_2^t}.$$

   The parameter update rule is:

   $$\theta := \theta - \frac{\eta}{\sqrt{\hat{v}_t} + \epsilon}\hat{m}_t.$$

Bias-correction is applied to counteract the shift towards zero of m_t and v_t at initialization, ensuring unbiased estimates. β_1 and β_2 are exponentially decaying averages that lead to effective adaptation in learning rates across iterations.

5. **Calculate the velocity term v after two iterations using momentum with $\beta = 0.9$, given initial velocity $v_0 = 0$, learning rate $\eta = 0.01$, and gradients $\nabla J(\theta)$ in the first and second iteration are -3 and -2, respectively.**

Solution: Iteration 1:

$$\mathbf{v}_1 = 0.9 \times 0 - 0.01 \times (-3) = 0.03$$

Iteration 2:

$$\mathbf{v}_2 = 0.9 \times 0.03 - 0.01 \times (-2) = 0.027 + 0.02 = 0.047$$

Thus, $\mathbf{v}$ after two iterations is 0.047.

6. **For the following sequence of gradients $[2, 4, 6]$, calculate the adjusted learning rates for each iteration using Adagrad with an initial learning rate $\eta = 0.01$ and $\epsilon = 10^{-8}$.**

Solution: Iteration 1:

$$G_1 = 2^2 = 4, \quad \text{Adjusted rate} = \frac{0.01}{\sqrt{4 + 10^{-8}}} = 0.005$$

Iteration 2:

$$G_2 = 4^2 + G_1 = 16 + 4 = 20, \quad \text{Adjusted rate} = \frac{0.01}{\sqrt{20 + 10^{-8}}} \approx 0.00224$$

Iteration 3:

$$G_3 = 6^2 + G_2 = 36 + 20 = 56, \quad \text{Adjusted rate} = \frac{0.01}{\sqrt{56 + 10^{-8}}} \approx 0.00133$$

Consequently, the adjusted learning rates for iterations 1, 2, and 3 are approximately 0.005, 0.00224, and 0.00133, respectively.

Practice Problems 3

1. Consider the momentum update rule for the parameter θ:

$$\mathbf{v} := \beta \mathbf{v} - \eta \nabla J(\theta), \quad \theta := \theta + \mathbf{v}$$

Assume $\beta = 0.9$, $\eta = 0.01$, initial values $\theta = 1$, and $\mathbf{v} = 0$. If $\nabla J(\theta) = 2$, compute the update for θ after one iteration.

2. Compute the velocity and parameter update using Nesterov Accelerated Gradient, given:

$$\mathbf{v} := \beta\mathbf{v} - \eta\nabla J(\theta + \beta\mathbf{v}), \quad \theta := \theta + \mathbf{v}$$

Assume $\beta = 0.9$, $\eta = 0.01$, $\theta = 1$, $\mathbf{v} = 0$, and $\nabla J(\theta) = 2$. What is the new value of θ?

3. Explore the Adagrad update mechanism for a parameter θ_i with:

$$\theta_i := \theta_i - \frac{\eta}{\sqrt{G_{ii} + \epsilon}}\nabla_{\theta_i} J(\theta)$$

Given $\eta = 0.01$, $G_{ii} = 4$, $\epsilon = 1e - 8$, and $\nabla_{\theta_i} J(\theta) = 3$, calculate the update for θ_i.

4. Calculate the parameter update for RMSprop using:

$$\mathbf{E}[g^2]_t := \gamma\mathbf{E}[g^2]_{t-1} + (1 - \gamma)(\nabla_{\theta_i} J(\theta))^2, \quad \theta_i := \theta_i - \frac{\eta}{\sqrt{\mathbf{E}[g^2]_t + \epsilon}}\nabla_{\theta_i} J(\theta)$$

Assume $\gamma = 0.9$, $\eta = 0.01$, $\epsilon = 1e - 8$, initial $\mathbf{E}[g^2]_{t-1} = 1$, and $\nabla_{\theta_i} J(\theta) = -5$.

5. Using Adam's optimizer, determine the new value of θ given:

$$m_t := \beta_1 m_{t-1} + (1 - \beta_1)\nabla J(\theta), \quad v_t := \beta_2 v_{t-1} + (1 - \beta_2)(\nabla J(\theta))^2$$

$$\hat{m}_t := \frac{m_t}{1 - \beta_1^t}, \quad \hat{v}_t := \frac{v_t}{1 - \beta_2^t}, \quad \theta := \theta - \frac{\eta}{\sqrt{\hat{v}_t} + \epsilon}\hat{m}_t$$

Given $\beta_1 = 0.9$, $\beta_2 = 0.999$, $\eta = 0.01$, $\epsilon = 1e - 8$, $\nabla J(\theta) = 4$, initial $m_{t-1} = 0$, $v_{t-1} = 0$, and $\theta = 3$.

6. Justify the importance of using adaptive learning rate algorithms in training deep neural networks, specifically focusing on the challenges addressed by them.

Answers 3

1. For the momentum update, with initial conditions $\theta = 1$, $\mathbf{v} = 0$, and $\nabla J(\theta) = 2$:

$$\mathbf{v} := 0.9 \times 0 - 0.01 \times 2 = -0.02$$

$$\theta := 1 + (-0.02) = 0.98$$

Therefore, the updated value of θ is 0.98.

2. For Nesterov Accelerated Gradient:

$$\mathbf{v} := 0.9 \times 0 - 0.01 \times 2 = -0.02$$

Then, calculate $\nabla J(\theta + \beta \mathbf{v}) \approx \nabla J(1 + 0.9 \times 0) = 2$.

$$\theta := 1 + (-0.02) = 0.98$$

Hence, the new value of θ is 0.98.

3. For Adagrad:

$$\theta_i := \theta_i - \frac{0.01}{\sqrt{4 + 1e-8}} \times 3 = \theta_i - \frac{0.01}{2.000000005} \times 3$$

$$\theta_i := \theta_i - 0.005 \approx \theta_i - 0.015$$

The parameter update θ_i is modified by -0.015.

4. For RMSprop:

$$\mathbf{E}[g^2]_t := 0.9 \times 1 + 0.1 \times (-5)^2 = 0.9 + 2.5 = 3.4$$

$$\theta_i := \theta_i - \frac{0.01}{\sqrt{3.4 + 1e-8}}(-5)$$

$$= \theta_i + \frac{0.01 \times 5}{1.84391} \approx \theta_i + 0.02714$$

The update to θ_i is approximately $+0.02714$.

5. Applying Adam's method:

$$m_t := 0.9 \times 0 + 0.1 \times 4 = 0.4$$

$$v_t := 0.999 \times 0 + 0.001 \times (4)^2 = 0.016$$

$$\hat{m}_t = \frac{0.4}{1 - 0.9^1} = 4, \quad \hat{v}_t = \frac{0.016}{1 - 0.999^1} = 16$$

$$\theta := 3 - \frac{0.01}{\sqrt{16 + 1e-8}} \times 4 = 3 - 0.01 \approx 2.99$$

The new θ value is approximately 2.99.

6. Adaptive learning rates are crucial in deep networks due to the following reasons:

- They allow each parameter to be individually adjusted, improving convergence and performance in networks with sparse gradients or varying correlation strengths across features.

- These methods mitigate the issue of selecting an optimal global learning rate, thereby enhancing training dynamics and avoiding pitfalls related to exploding or vanishing gradient problems.

Chapter 30

Batch Normalization and Calculus

Practice Problems 1

1. Calculate the gradient of the affine transformation output y_i with respect to the normalized inputs $\hat{x}_i$.

$$y_i = \gamma \hat{x}_i + \beta$$

2. Derive the expression for the gradient of the batch variance σ^2 with respect to the loss $\mathcal{L}$.

$$\text{Use: } \sigma^2 = \frac{1}{m} \sum_{i=1}^{m} (x_i - \mu)^2$$

3. Compute the gradient of the mini-batch mean μ with respect to the loss $\mathcal{L}$.

$$\text{Given: } \mu = \frac{1}{m} \sum_{i=1}^{m} x_i$$

4. Derive the gradient of the loss function with respect to each input x_i considering the interdependency with mean and variance.

$$\text{Given: } \hat{x}_i = \frac{x_i - \mu}{\sqrt{\sigma^2 + \epsilon}}$$

5. Explain how batch normalization can regularize a neural network and mention any potential downside of this technique.

6. Describe the role of the small constant ϵ in the normalization process and its significance in computational stability.

Answers 1

1. Calculate the gradient of the affine transformation output y_i with respect to the normalized inputs $\hat{x}_i$.

   Solution:

   $$y_i = \gamma \hat{x}_i + \beta$$

   The gradient of y_i with respect to $\hat{x}_i$ is simply the partial derivative:

   $$\frac{\partial y_i}{\partial \hat{x}_i} = \gamma$$

   Therefore, the gradient is:

   $$\frac{\partial y_i}{\partial \hat{x}_i} = \gamma.$$

284

2. Derive the expression for the gradient of the batch variance σ^2 with respect to the loss $\mathcal{L}$.

Solution:

$$\sigma^2 = \frac{1}{m} \sum_{i=1}^{m} (x_i - \mu)^2$$

Using the law of total derivatives, the gradient of the loss with respect to σ^2 is:

$$\frac{\partial \mathcal{L}}{\partial \sigma^2} = \sum_{i=1}^{m} \frac{\partial \mathcal{L}}{\partial \hat{x}_i} \cdot \frac{\partial \hat{x}_i}{\partial \sigma^2}$$

$$= \sum_{i=1}^{m} \frac{\partial \mathcal{L}}{\partial \hat{x}_i} \cdot (x_i - \mu) \cdot \left(-\frac{1}{2} \right) \cdot (\sigma^2 + \epsilon)^{-3/2}$$

Therefore, the gradient is:

$$\frac{\partial \mathcal{L}}{\partial \sigma^2} = \sum_{i=1}^{m} \frac{\partial \mathcal{L}}{\partial \hat{x}_i} \cdot (x_i - \mu) \cdot (-\frac{1}{2}) \cdot (\sigma^2 + \epsilon)^{-\frac{3}{2}}.$$

3. Compute the gradient of the mini-batch mean μ with respect to the loss $\mathcal{L}$.

Solution:

$$\mu = \frac{1}{m} \sum_{i=1}^{m} x_i$$

The gradient of the loss with respect to μ is:

$$\frac{\partial \mathcal{L}}{\partial \mu} = \sum_{i=1}^{m} \frac{\partial \mathcal{L}}{\partial \hat{x}_i} \cdot \left(-\frac{1}{\sqrt{\sigma^2 + \epsilon}} \right) + \frac{\partial \mathcal{L}}{\partial \sigma^2} \cdot \left(-\frac{2}{m} \right) \sum_{i=1}^{m} (x_i - \mu)$$

Therefore, the gradient is:

$$\frac{\partial \mathcal{L}}{\partial \mu} = \sum_{i=1}^{m} \frac{\partial \mathcal{L}}{\partial \hat{x}_i} \cdot (-\frac{1}{\sqrt{\sigma^2 + \epsilon}}) + \frac{\partial \mathcal{L}}{\partial \sigma^2} \cdot (-\frac{2}{m}) \sum_{i=1}^{m} (x_i - \mu).$$

4. Derive the gradient of the loss function with respect to each input x_i.

Solution:

$$\hat{x}_i = \frac{x_i - \mu}{\sqrt{\sigma^2 + \epsilon}}$$

The derivative of the loss function with respect to each input x_i is:

$$\frac{\partial \mathcal{L}}{\partial x_i} = \frac{\partial \mathcal{L}}{\partial \hat{x}_i} \cdot \frac{1}{\sqrt{\sigma^2 + \epsilon}} + \frac{\partial \mathcal{L}}{\partial \sigma^2} \cdot \frac{2(x_i - \mu)}{m} + \frac{\partial \mathcal{L}}{\partial \mu} \cdot \frac{1}{m}$$

Therefore, the gradient is:

$$\frac{\partial \mathcal{L}}{\partial x_i} = \frac{\partial \mathcal{L}}{\partial \hat{x}_i} \cdot \frac{1}{\sqrt{\sigma^2 + \epsilon}} + \frac{\partial \mathcal{L}}{\partial \sigma^2} \cdot \frac{2(x_i - \mu)}{m} + \frac{\partial \mathcal{L}}{\partial \mu} \cdot \frac{1}{m}.$$

5. Explain how batch normalization can regularize a neural network and mention any potential downside of this technique.

Solution: Batch normalization adds noise to layer inputs during training due to variations in batches, acting as a regularizer. It mitigates internal covariate shift, leading to faster convergence and allowing the use of higher learning rates. Potential downsides include computational overhead from additional calculations of mean and variance for each mini-batch, which may reduce training speed.

6. Describe the role of the small constant ϵ in the normalization process and its significance in computational stability.

   Solution: ϵ is introduced to prevent division by zero or extremely small numbers during normalization:

   $$\hat{x}_i = \frac{x_i - \mu}{\sqrt{\sigma^2 + \epsilon}}$$

   This small constant ensures numerical stability by keeping the denominator non-zero. Without ϵ, variability in the variance, especially when close to zero, could result in infinite or undefined behaviors during computation.

Practice Problems 2

1. Derive the expression for the gradient of the loss function with respect to the affine transformation parameter γ in batch normalization.

2. Prove how the derivative of the loss function with respect to the mean μ is obtained in batch normalization.

3. Calculate the derivative of the normalized input $\hat{x}_i$ concerning the original input x_i.

4. Explain the significance of the ϵ term in the variance normalization step in batch normalization.

5. Analyze how batch normalization affects the gradient flow within the network layers and discuss its impact on convergence.

6. Verify the expression for the gradient of the input x_i derived from batch normalization, using the chain rule.

Answers 2

1. Derive the expression for the gradient of the loss function with respect to the affine transformation parameter γ.

   Solution:

   The affine transformation in batch normalization is given by:

   $$y_i = \gamma \hat{x}_i + \beta$$

   During backpropagation, we need $\frac{\partial \mathcal{L}}{\partial \gamma}$, where $\mathcal{L}$ is the loss function. Using the chain rule, it can be expressed as:

   $$\frac{\partial \mathcal{L}}{\partial \gamma} = \sum_{i=1}^{m} \frac{\partial \mathcal{L}}{\partial y_i} \cdot \hat{x}_i$$

   This indicates that for each element of the batch, the derivative of the loss with respect to γ is the weighted sum of normalized inputs.

2. Prove how the derivative of the loss function with respect to the mean μ is obtained in batch normalization.

   Solution:

   The partial derivative of the loss with respect to the mean is expressed as:

   $$\frac{\partial \mathcal{L}}{\partial \mu} = \sum_{i=1}^{m} \frac{\partial \mathcal{L}}{\partial \hat{x}_i} \cdot \left(-\frac{1}{\sqrt{\sigma^2 + \epsilon}} \right) + \frac{\partial \mathcal{L}}{\partial \sigma^2} \cdot \left(-\frac{2}{m} \right) \sum_{i=1}^{m} (x_i - \mu)$$

   This is derived by applying total derivatives, as μ affects both the normalized inputs and indirectly the variance through the input terms $x_i - \mu$.

3. Calculate the derivative of the normalized input $\hat{x}_i$ concerning the original input x_i.

   Solution:

   The normalized input is given as:
   $$\hat{x}_i = \frac{x_i - \mu}{\sqrt{\sigma^2 + \epsilon}}$$

   Applying the chain rule, the derivative with respect to x_i is:
   $$\frac{\partial \hat{x}_i}{\partial x_i} = \frac{1}{\sqrt{\sigma^2 + \epsilon}} - \frac{(x_i - \mu)}{m(\sigma^2 + \epsilon)^{3/2}}$$

   This expression shows how each x_i contributes to the normalized input taking into account both its deviation from the batch mean and the overall batch variance.

4. Explain the significance of the ϵ term in the variance normalization step in batch normalization.

   Solution:

   The term ϵ is crucial for numerical stability. In batch normalization:
   $$\hat{x}_i = \frac{x_i - \mu}{\sqrt{\sigma^2 + \epsilon}}$$

   If σ^2 is very small, the division could become undefined or extremely large. The ϵ term prevents division by zero and ensures stable computation of gradients by providing a lower bound to the variance.

5. Analyze how batch normalization affects the gradient flow within the network layers and discuss its impact on convergence.

   Solution:

   Batch normalization standardizes the inputs to each layer to have mean zero and variance one, thus regularizing the data distribution across layers. This reduces the internal covariate shift, which refers to changes in the distribution of network activations due to parameter updates. By controlling this shift, gradient flow is stabilized, helping the network converge faster and potentially enabling higher learning rates without risk of divergence.

6. Verify the expression for the gradient of the input x_i derived from batch normalization using the chain rule.

   Solution:

   We want to verify:
   $$\frac{\partial \mathcal{L}}{\partial x_i} = \frac{\partial \mathcal{L}}{\partial \hat{x}_i} \cdot \frac{1}{\sqrt{\sigma^2 + \epsilon}} + \frac{\partial \mathcal{L}}{\partial \sigma^2} \cdot \frac{2(x_i - \mu)}{m} + \frac{\partial \mathcal{L}}{\partial \mu} \cdot \frac{1}{m}$$

   Using the chain and product rules, this is obtained by first noting the contribution of x_i to $\hat{x}_i$, σ^2, and μ. Each of these dependencies is accounted for, ensuring that changes in x_i are propagated accurately into the final loss gradient. This ensures that all relationships through $\hat{x}_i$ and the batch statistics are captured within updates to x_i.

Practice Problems 3

1. If a mini-batch of inputs is given as $\mathbf{x} = [2, 4, 6, 8]$, compute the normalized inputs $\hat{x}_i$ using batch normalization.

2. For the given mini-batch in problem 1, calculate the batch-normalized outputs y_i using $\gamma = 2$ and $\beta = 1$.

3. Derive the expression for $\frac{\partial \mathcal{L}}{\partial \gamma}$ for a single mini-batch where $\mathcal{L}$ is the loss function.

4. Consider a case where $\hat{x}_i$ causes the gradients to explode. Briefly explain why batch normalization can help mitigate this issue.

5. Compute the derivative $\frac{\partial \mathcal{L}}{\partial \beta}$ for a loss function $\mathcal{L}$ with respect to β in a batch normalization layer.

6. Explain the role of ϵ in the normalization step and its impact on numerical stability.

Answers 3

1. **Solution:** For the mini-batch $\mathbf{x} = [2, 4, 6, 8]$, compute μ and σ^2:

$$\mu = \frac{1}{4}(2 + 4 + 6 + 8) = 5$$

$$\sigma^2 = \frac{1}{4}((2 - 5)^2 + (4 - 5)^2 + (6 - 5)^2 + (8 - 5)^2) = \frac{1}{4}(9 + 1 + 1 + 9) = 5$$

Then, the normalized inputs $\hat{x}_i$:

$$\hat{x}_i = \frac{x_i - \mu}{\sqrt{\sigma^2 + \epsilon}}$$

Assuming $\epsilon = 0$:

$$\hat{x}_1 = \frac{2 - 5}{\sqrt{5}} = -\frac{3}{\sqrt{5}}, \quad \hat{x}_2 = \frac{4 - 5}{\sqrt{5}} = -\frac{1}{\sqrt{5}}$$

$$\hat{x}_3 = \frac{6 - 5}{\sqrt{5}} = \frac{1}{\sqrt{5}}, \quad \hat{x}_4 = \frac{8 - 5}{\sqrt{5}} = \frac{3}{\sqrt{5}}$$

2. **Solution:** Using $\gamma = 2$ and $\beta = 1$:

$$y_i = \gamma \hat{x}_i + \beta$$

Compute each y_i:

$$y_1 = 2\left(-\frac{3}{\sqrt{5}}\right) + 1 = 1 - \frac{6}{\sqrt{5}}, \quad y_2 = 2\left(-\frac{1}{\sqrt{5}}\right) + 1 = 1 - \frac{2}{\sqrt{5}}$$

$$y_3 = 2\left(\frac{1}{\sqrt{5}}\right) + 1 = 1 + \frac{2}{\sqrt{5}}, \quad y_4 = 2\left(\frac{3}{\sqrt{5}}\right) + 1 = 1 + \frac{6}{\sqrt{5}}$$

3. **Solution:** Given $\hat{x}_i$, the gradient $\frac{\partial \mathcal{L}}{\partial \gamma}$ can be deduced as:

$$\frac{\partial \mathcal{L}}{\partial \gamma} = \sum_{i=1}^{m} \frac{\partial \mathcal{L}}{\partial y_i} \cdot \hat{x}_i$$

This expression accounts for how much the loss function $\mathcal{L}$ changes with respect to the scaling factor γ.

4. **Solution:** In backpropagation, gradients can become large (exploding) or small (vanishing), which batch normalization helps mitigate by normalizing intermediate outputs. By ensuring each layer's output is normalized, it constrains the input distribution, maintaining a stable gradient trajectory for learning.

5. **Solution:** The derivative $\frac{\partial \mathcal{L}}{\partial \beta}$ for the loss function $\mathcal{L}$ can be computed as:

$$\frac{\partial \mathcal{L}}{\partial \beta} = \sum_{i=1}^{m} \frac{\partial \mathcal{L}}{\partial y_i}$$

This represents how the shift parameter β contributes to the networks output discrepancy represented by the loss $\mathcal{L}$.

6. **Solution:** The term ϵ is a small constant added to the variance σ^2 to prevent division by zero during normalization. It ensures numerical stability, especially when variance is close to zero, avoiding large spikes in normalized values and, consequently, stabilizing training.

Chapter 31

Calculus in Support Vector Machines

Practice Problems 1

1. Explain how the primal optimization problem of a Support Vector Machine (SVM) ensures the maximization of the margin. What role does the weight vector **w** play in this context?

2. Derive the dual problem formulation from the primal problem of an SVM. Explain how the constraints in the primal problem lead to the dual problem's objective and constraints.

3. In the context of SVMs, explain the significance of Lagrange multipliers in transitioning from the primal to the dual problem.

4. Describe the kernel trick for SVMs and explain how it enables dealing with non-linear data separations. Provide an example of a commonly used kernel function.

5. How can the soft margin approach in SVMs handle non-separability in data? What is the role of slack variables, and how does this affect the objective function?

6. Consider the dual problem of an SVM. Show how the support vectors are identified and explain their role in the decision boundary.

Answers 1

1. **Solution:** The primal optimization problem of an SVM is to minimize $\frac{1}{2}\|\mathbf{w}\|^2$ subject to the constraint $y_i(\mathbf{w} \cdot \mathbf{x}_i + b) \geq 1$. This optimization ensures that the hyperplane maximizes the separation (or margin) between classes because minimizing $\frac{1}{2}\|\mathbf{w}\|^2$ is equivalent to maximizing the distance from the hyperplane to the nearest data point from each class. The weight vector $\mathbf{w}$ determines the orientation and slope of the separating hyperplane. A smaller norm of $\mathbf{w}$ indicates a larger margin.

2. **Solution:** To derive the dual problem, we start with the primal problem:

$$\text{minimize:} \quad \frac{1}{2}\|\mathbf{w}\|^2$$

$$\text{subject to:} \quad y_i(\mathbf{w} \cdot \mathbf{x}_i + b) \geq 1$$

We introduce Lagrange multipliers $\alpha_i \geq 0$ for these constraints and form the Lagrangian:

$$\mathcal{L}(\mathbf{w}, b, \boldsymbol{\alpha}) = \frac{1}{2}\|\mathbf{w}\|^2 - \sum_{i=1}^{m} \alpha_i[y_i(\mathbf{w} \cdot \mathbf{x}_i + b) - 1]$$

By setting the partial derivatives of $\mathcal{L}$ with respect to $\mathbf{w}$ and b to zero, we obtain constraints that allow α_i to form the dual problem:

$$\mathbf{w} = \sum_{i=1}^{m} \alpha_i y_i \mathbf{x}_i, \quad \sum_{i=1}^{m} \alpha_i y_i = 0$$

Substituting these back gives us the dual objective:

$$\text{maximize:} \quad \sum_{i=1}^{m} \alpha_i - \frac{1}{2} \sum_{i=1}^{m} \sum_{j=1}^{m} \alpha_i \alpha_j y_i y_j \mathbf{x}_i \cdot \mathbf{x}_j$$

with constraints $\alpha_i \geq 0$.

3. **Solution:** Lagrange multipliers α_i are introduced to handle constraints in the primal optimization problem. They allow the conversion of inequality constraints into conditions that contribute to the Lagrangian. By optimizing the Lagrangian, we effectively solve for the conditions under which the primal problem holds and derive a dual problem that is often easier to solve. Each non-zero α_i in the optimal solution indicates that the corresponding constraint is active at the boundary, identifying data points that are support vectors constructing the decision boundary.

4. **Solution:** The kernel trick allows SVMs to solve non-linear problems by mapping input features into higher-dimensional feature spaces, where linear separation is possible. Instead of explicitly performing this transformation $\phi(\mathbf{x})$, we use a kernel function $K(\mathbf{x}_i, \mathbf{x}_j) = \phi(\mathbf{x}_i) \cdot \phi(\mathbf{x}_j)$, which computes the inner product directly in the transformed space. A popular kernel is the Radial Basis Function (RBF) kernel:

$$K(\mathbf{x}_i, \mathbf{x}_j) = \exp\left(-\frac{\|\mathbf{x}_i - \mathbf{x}_j\|^2}{2\sigma^2}\right)$$

5. **Solution:** The soft margin approach introduces slack variables ξ_i to allow some misclassifications. The constraints are modified to:

$$y_i(\mathbf{w} \cdot \mathbf{x}_i + b) \geq 1 - \xi_i, \quad \xi_i \geq 0$$

The new objective function becomes:

$$\frac{1}{2}\|\mathbf{w}\|^2 + C \sum_{i=1}^{m} \xi_i$$

where C is the regularization parameter that balances maximizing margin width and minimizing classification error. Slack variables allow the margin boundary to relax to accommodate data violations.

6. **Solution:** In the dual-formulation optimization, the resulting solution includes non-zero Lagrange multipliers α_i associated with specific data points called support vectors. These vectors are crucial as they lie closest to the hyperplane and bear on its position and orientation. The decision boundary depends entirely on these support vectors, formalized in the classifier function:

$$f(\mathbf{x}) = \sum_{i=1}^{m} \alpha_i y_i K(\mathbf{x}_i, \mathbf{x}) + b$$

Thus, $\alpha_i > 0$ indicates which training points $\mathbf{x}_i$ are support vectors and contribute to defining the hyperplane.

Practice Problems 2

1. Derive the dual formulation from the primal optimization problem in Support Vector Machines:

$$\text{Minimize:} \quad \frac{1}{2}\|\mathbf{w}\|^2 \quad \text{subject to:} \quad y_i(\mathbf{w} \cdot \mathbf{x}_i + b) \geq 1, \quad \forall i$$

2. Given a set of support vectors, how do you determine the bias term b in the decision function?

3. Explain the role of the kernel function in transforming a Support Vector Machine and provide an example of a common kernel.

4. Analyze the impact of the regularization parameter C on the soft margin SVM and describe its influence on the model complexity.

5. What are the partial derivatives necessary for the gradient descent algorithm on the dual problem of SVM?

6. Derive the expression for a decision function $f(\mathbf{x})$ of an SVM using support vectors and Lagrange multipliers.

Answers 2

1. Derive the dual formulation:

The primal optimization problem is: minimize: $\quad \dfrac{1}{2}\|\mathbf{w}\|^2$

subject to: $\quad y_i(\mathbf{w} \cdot \mathbf{x}_i + b) \geq 1, \quad \forall i$

Introducing Lagrange multipliers $\alpha_i \geq 0$ for constraints, we form the Lagrangian:

$$\mathcal{L}(\mathbf{w}, b, \boldsymbol{\alpha}) = \frac{1}{2}\|\mathbf{w}\|^2 - \sum_{i=1}^{m} \alpha_i[y_i(\mathbf{w} \cdot \mathbf{x}_i + b) - 1]$$

Taking derivatives with respect to $\mathbf{w}$ and b and setting them to zero:

$$\frac{\partial \mathcal{L}}{\partial \mathbf{w}} = \mathbf{w} - \sum_{i=1}^{m} \alpha_i y_i \mathbf{x}_i = 0 \quad \Rightarrow \quad \mathbf{w} = \sum_{i=1}^{m} \alpha_i y_i \mathbf{x}_i$$

$$\frac{\partial \mathcal{L}}{\partial b} = -\sum_{i=1}^{m} \alpha_i y_i = 0$$

Substituting back, the dual problem becomes:

$$\text{maximize:} \quad \sum_{i=1}^{m} \alpha_i - \frac{1}{2}\sum_{i=1}^{m}\sum_{j=1}^{m} \alpha_i \alpha_j y_i y_j \mathbf{x}_i \cdot \mathbf{x}_j$$

$$\text{subject to:} \quad \sum_{i=1}^{m} \alpha_i y_i = 0, \quad \alpha_i \geq 0 \quad \forall i$$

2. Determine the bias term b:

After solving the dual problem and obtaining the support vectors, pick any support vector $\mathbf{x}_s$ for which $0 < \alpha_s < C$. Use:

$$y_s(\mathbf{w} \cdot \mathbf{x}_s + b) = 1$$

Substituting $\mathbf{w}$ from previously calculated support vectors:

$$b = y_s - \sum_{i=1}^{m} \alpha_i y_i (\mathbf{x}_i \cdot \mathbf{x}_s)$$

3. **Role of kernel function:**

   The kernel function $K(\mathbf{x}_i, \mathbf{x}_j)$ allows SVMs to create a decision boundary in a transformed feature space without explicitly computing coordinates in that space, termed as the "kernel trick":

   $$K(\mathbf{x}_i, \mathbf{x}_j) = \phi(\mathbf{x}_i) \cdot \phi(\mathbf{x}_j)$$

   A common kernel function example is the Radial Basis Function (RBF) kernel:

   $$K(\mathbf{x}_i, \mathbf{x}_j) = \exp\left(-\gamma \|\mathbf{x}_i - \mathbf{x}_j\|^2\right)$$

4. **Impact of regularization parameter C:**

   The parameter C controls the trade-off between achieving a large margin and minimizing classification error. A small C encourages a larger margin, potentially allowing some misclassifications, whereas a large C aims at classifying all training samples correctly:

   - Smaller C: Larger margin, less complex model, greater tolerance to errors.
   - Larger C: Smaller margin, more complex model, less tolerance to errors.

5. **Partial derivatives for gradient descent:**

   The derivative of the dual objective with respect to α_i:

   $$\frac{\partial \mathcal{L}_{dual}}{\partial \alpha_i} = 1 - y_i \left(\sum_{j=1}^{m} \alpha_j y_j K(\mathbf{x}_j, \mathbf{x}_i) \right)$$

   These derivatives guide updates for each α_i during gradient descent.

6. **Derive the decision function:**

   Given the support vectors and the Lagrange multipliers from the dual solution, the decision function for a given data point $\mathbf{x}$ is:

   $$f(\mathbf{x}) = \sum_{i=1}^{m} \alpha_i y_i K(\mathbf{x}_i, \mathbf{x}) + b$$

   Here, the support vectors contribute to defining the hyperplane decision boundary, and the bias term b shifts the hyperplane in space.

Practice Problems 3

1. Consider the following optimization problem. Minimize:

   $$\frac{1}{2} \|\mathbf{w}\|^2$$

   Subject to:

   $$y_i(\mathbf{w} \cdot \mathbf{x}_i + b) \geq 1$$

   Write down the Lagrangian and explain the steps to convert it into the dual problem.

2. For a given SVM with Lagrange multipliers α_i, derive the condition under which a data point $\mathbf{x}_i$ is a support vector.

3. Given a kernel function $K(\mathbf{x}, \mathbf{z}) = (\mathbf{x} \cdot \mathbf{z})^2$, express the transformed feature space representation and describe the benefits of applying such a transformation in an SVM.

4. Demonstrate how the soft margin SVM formulation alleviates issues found in hard margin SVM when dealing with non-linearly separable data. Provide illustrative explanations.

5. Calculate the gradient of the dual objective function with respect to the Lagrange multipliers α_i for a given SVM and discuss the significance of the gradient in the optimization process.

6. If a new data point $\mathbf{z}$ is introduced in a trained SVM model, describe the method to update the model to include this new data point without retraining from scratch. Consider the computational aspects.

Answers 3

1. **Solution:** The Lagrangian for the problem can be written as:

$$\mathcal{L}(\mathbf{w}, b, \boldsymbol{\alpha}) = \frac{1}{2}\|\mathbf{w}\|^2 - \sum_{i=1}^{m} \alpha_i[y_i(\mathbf{w} \cdot \mathbf{x}_i + b) - 1]$$

To convert to the dual, we set the gradients of $\mathcal{L}$ with respect to $\mathbf{w}$ and b to zero:

$$\frac{\partial \mathcal{L}}{\partial \mathbf{w}} = \mathbf{w} - \sum_{i=1}^{m} \alpha_i y_i \mathbf{x}_i = 0$$

$$\frac{\partial \mathcal{L}}{\partial b} = -\sum_{i=1}^{m} \alpha_i y_i = 0$$

The dual problem becomes:

$$\text{maximize:} \quad \sum_{i=1}^{m} \alpha_i - \frac{1}{2}\sum_{i=1}^{m}\sum_{j=1}^{m} \alpha_i \alpha_j y_i y_j \mathbf{x}_i \cdot \mathbf{x}_j$$

Subject to:

$$\alpha_i \geq 0, \quad \sum_{i=1}^{m} \alpha_i y_i = 0$$

2. **Solution:** A data point $\mathbf{x}_i$ is a support vector if its corresponding Lagrange multiplier α_i is non-zero. Specifically, this happens when:
$$y_i(\mathbf{w} \cdot \mathbf{x}_i + b) = 1$$
Support vectors lie exactly on the margin, where they fulfill the equality condition in the constraint.

3. **Solution:** The kernel function $K(\mathbf{x}, \mathbf{z}) = (\mathbf{x} \cdot \mathbf{z})^2$ corresponds to a feature mapping:
$$K(\mathbf{x}, \mathbf{z}) = \phi(\mathbf{x}) \cdot \phi(\mathbf{z})$$

This maps the data implicitly into a higher-dimensional space, allowing for non-linear separation while retaining computational efficiency due to the kernel trick. The benefit is that it allows the SVM to construct more complex decision boundaries without explicitly transforming the data.

4. **Solution:** The soft margin SVM allows for some violations of the margin constraints by introducing slack variables ξ_i:
$$y_i(\mathbf{w} \cdot \mathbf{x}_i + b) \geq 1 - \xi_i, \quad \xi_i \geq 0$$

This formulation permits some misclassifications, which enhances the model's ability to generalize by avoiding overfitting to noisy or outlier data. The regularization parameter C controls the trade-off between maximizing the margin and minimizing classification error.

5. **Solution:** The gradient of the dual objective function with respect to α_i is:

$$\frac{\partial \mathcal{L}_D}{\partial \alpha_i} = 1 - \sum_{j=1}^{m} \alpha_j y_i y_j \mathbf{x}_i \cdot \mathbf{x}_j$$

The gradient provides the direction in which the dual objective function should be adjusted to reach the maximum. It is crucial for iterative optimization methods such as gradient ascent.

6. **Solution:** When a new data point $\mathbf{z}$ is introduced, the SVM model can be updated by revisiting the current optimization problem and slightly modifying it to include the term associated with $\mathbf{z}$. This involves updating the Gram matrix with the kernel evaluations of $\mathbf{z}$ and re-optimizing the primal/dual formulations. Techniques such as "warm-starting" the optimization with the previous solution can save computational resources by using the existing model parameters as a starting point.

Chapter 32

Calculus in Decision Tree Algorithms

Practice Problems 1

1. Compute the impurity decrease given a node with Gini impurity 0.5 is split into two child nodes with impurities 0.3 and 0.25, and the left child contains 60% of the instances. Calculate the impurity decrease.

2. Consider a decision tree boosting scenario where the initial model prediction $\hat{y}_i$ and true label y_i are known. If the loss function is the squared error $L(y, \hat{y}) = (y - \hat{y})^2$, derive the expression for the pseudo-residual.

3. Show that the entropy-based impurity measure reaches its maximum when classes are equally probable. Use two classes for simplicity in your solution.

4. A gradient boosting process uses a particular learning rate η. Explain the role of this learning rate in convergence speed and model performance, considering overfitting.

5. Discuss the role of second-order derivatives in XGBoost. Explain how they can contribute to faster convergence compared to using just first-order derivatives.

6. For a binary classification problem, consider the gradient boosting framework. Derive the update rule for model weights using the logistic loss function $L(y, \hat{y}) = -y \log(\hat{y}) - (1 - y) \log(1 - \hat{y})$.

Answers 1

1. Compute the impurity decrease given a node with Gini impurity 0.5 is split into two child nodes with impurities 0.3 and 0.25, and the left child contains 60% of the instances. Calculate the impurity decrease.

   Solution:
   The impurity decrease ΔI is calculated as:

   $$\Delta I = I(\text{parent}) - [p \times I(\text{left child}) + (1 - p) \times I(\text{right child})]$$

   where $p = 0.6$ is the proportion of instances in the left child. Substituting the values, we have:

   $$\Delta I = 0.5 - [0.6 \times 0.3 + 0.4 \times 0.25]$$

   $$= 0.5 - (0.18 + 0.1) = 0.5 - 0.28 = 0.22$$

   Therefore, the impurity decrease is 0.22.

2. Consider a decision tree boosting scenario where the initial model prediction $\hat{y}_i$ and true label y_i are known. If the loss function is the squared error $L(y, \hat{y}) = (y - \hat{y})^2$, derive the expression for the pseudo-residual.

Solution:

The pseudo-residual is the negative gradient of the loss function concerning the current prediction $\hat{y}_i$:

$$r_i = -\frac{\partial L(y_i, \hat{y}_i)}{\partial \hat{y}_i} = -2(y_i - \hat{y}_i)(-1) = 2(y_i - \hat{y}_i)$$

Therefore, the pseudo-residual for squared error is $2(y_i - \hat{y}_i)$.

3. Show that the entropy-based impurity measure reaches its maximum when classes are equally probable. Use two classes for simplicity in your solution.

Solution:

The entropy for two classes is:

$$I(t) = -p_1 \log_2(p_1) - p_2 \log_2(p_2)$$

For maximum entropy, set $p_1 = p_2 = 0.5$, then:

$$I(t) = -0.5 \log_2(0.5) - 0.5 \log_2(0.5) = -0.5(-1) - 0.5(-1) = 1$$

Thus, entropy is maximized at 1 when both classes are equally probable.

4. A gradient boosting process uses a particular learning rate η. Explain the role of this learning rate in convergence speed and model performance, considering overfitting.

Solution:

The learning rate η determines the step size in the gradient descent direction during model updates. A smaller η slows convergence, allowing finer adjustments and potentially improving generalization by preventing overshooting or overfitting. A larger η speeds convergence but risks overfitting, as the model may converge to local minima quickly without thoroughly exploring the solution space.

5. Discuss the role of second-order derivatives in XGBoost. Explain how they can contribute to faster convergence compared to using just first-order derivatives.

Solution:

Second-order derivatives provide curvature information, which helps in adjusting the model weights more accurately in directions of steepest descent, facilitating faster convergence. In XGBoost, the Hessian (matrix of second derivatives) is used to make precise weight updates using a Taylor series approximation, enabling larger and more accurate update steps, reducing the number of iterations needed to reach optimal weights compared to using solely first-order derivatives.

6. For a binary classification problem, consider the gradient boosting framework. Derive the update rule for model weights using the logistic loss function $L(y, \hat{y}) = -y \log(\hat{y}) - (1 - y) \log(1 - \hat{y})$.

Solution:

Given the logistic loss:

$$L(y, \hat{y}) = -y \log(\hat{y}) - (1 - y) \log(1 - \hat{y})$$

The gradient with respect to $\hat{y}$ (consider $\hat{y} = p = \text{sigmoid}(f(x))$) is:

$$\nabla_{\hat{y}} L = -\frac{y}{\hat{y}} + \frac{1 - y}{1 - \hat{y}} = \hat{y} - y$$

For boosting, the update is:

$$\theta \leftarrow \theta - \eta(\hat{y} - y)$$

Where the update represents shifting the model prediction towards the true labels using the gradient of logistic loss.

Practice Problems 2

1. For a decision tree using Gini impurity, express how you would compute the reduction in impurity ΔI mathematically when a node splits into two child nodes.

2. If the impurity measure changes only by a small amount δ, use a first-order Taylor expansion to approximate how this change affects the impurity reduction ΔI during a node split.

3. Explain how the gradient descent method is utilized in training decision trees and describe what condition would indicate convergence of the tree parameters in a gradient boosting setup.

4. Derive the update step for a single parameter θ in a decision tree using a first-order gradient approximation for a given loss function $L(y, \hat{y})$.

5. How does the inclusion of second-order derivatives (e.g., Hessian) improve convergence in advanced boosting algorithms like XGBoost compared to traditional gradient descent?

6. Calculate the new model update in gradient boosting after fitting a regression tree to the residuals, given pseudo-residuals r_i and learning rate η.

Answers 2

1. For a decision tree using Gini impurity, express how you would compute the reduction in impurity ΔI mathematically when a node splits into two child nodes.

   Solution:

   The reduction in impurity ΔI when splitting a node t into child nodes t_L and t_R is given by:

   $$\Delta I = I(t) - \left(\frac{|t_L|}{|t|} I(t_L) + \frac{|t_R|}{|t|} I(t_R) \right)$$

   where $|t|$, $|t_L|$, and $|t_R|$ are the number of samples in nodes t, t_L, and t_R, respectively. $I(t)$, $I(t_L)$, and $I(t_R)$ are the impurity measures of these nodes.

2. If the impurity measure changes only by a small amount δ, use a first-order Taylor expansion to approximate how this change affects the impurity reduction ΔI during a node split.

   Solution:

   Using a first-order Taylor expansion, the change in impurity reduction can be approximated as:

   $$\Delta \tilde{I} = \frac{\partial I(t)}{\partial \delta} \delta - \left(\frac{|t_L|}{|t|} \frac{\partial I(t_L)}{\partial \delta} \delta + \frac{|t_R|}{|t|} \frac{\partial I(t_R)}{\partial \delta} \delta \right)$$

   This shows how small changes in impurity affect nodes proportionally to their derivatives with respect to δ.

3. Explain how the gradient descent method is utilized in training decision trees and describe what condition would indicate convergence of the tree parameters in a gradient boosting setup.

   Solution:

   Gradient descent is used in decision tree training to adjust parameters by iteratively minimizing a differentiable loss function. The update step is typically $\theta \leftarrow \theta - \eta \nabla_\theta L(y, \hat{y})$, where η is the learning rate. Convergence is indicated when changes in the parameters θ are smaller than a pre-defined tolerance level, signaling that further updates provide negligible improvement.

4. Derive the update step for a single parameter θ in a decision tree using a first-order gradient approximation for a given loss function $L(y, \hat{y})$.

   Solution:

   The parameter update using first-order gradient approximation follows:

   $$\theta \leftarrow \theta - \eta \nabla_\theta L(y, \hat{y})$$

   where $\nabla_\theta L(y, \hat{y})$ is the gradient of the loss function with respect to θ. This technique incrementally decreases the loss value by moving in the opposite direction of the gradient.

5. How does the inclusion of second-order derivatives (e.g., Hessian) improve convergence in advanced boosting algorithms like XGBoost compared to traditional gradient descent?

Solution:

The inclusion of second-order derivatives allows for utilizing the Hessian, which provides curvature information of the loss function. This leads to more accurate approximations of the loss surface and results in faster convergence rates since it allows for more informed updates of parameters. Specifically, the update step becomes:

$$\theta \leftarrow \theta - \eta \frac{\sum_i g_i}{\sum_i h_i + \lambda}$$

where g_i and h_i are the first and second derivatives respectively, improving efficiency by refining step direction and size.

6. Calculate the new model update in gradient boosting after fitting a regression tree to the residuals, given pseudo-residuals r_i and learning rate η.

Solution:

After calculating pseudo-residuals and fitting a regression tree $h_m(x)$ to these residuals, the model is updated as:

$$F(x) \leftarrow F(x) + \eta h_m(x)$$

Each tree added to the ensemble model learns patterns not captured by previous trees, and the learning rate η controls the contribution of each tree to prevent overfitting.

Practice Problems 3

1. Consider a decision tree where the impurity of a node t is defined by entropy. Derive the expression for the change in impurity ΔI when a node splits into two child nodes t_L and t_R.

2. Given the loss function $L(y_i, \hat{y}_i) = (y_i - \hat{y}_i)^2$, find the gradient used in gradient boosting for a node t.

3. In the context of gradient boosting, explain how second-order derivatives (Hessians) are utilized to update tree parameters more effectively.

4. Consider a dataset $\mathcal{D} = \{(x_i, y_i)\}_{i=1}^n$. Derive the expression to initialize the model $F(x)$ using a constant value that minimizes the loss $L(y_i, \gamma)$ over all the examples.

5. Explain how the chain rule in calculus is applied in the gradient boosting framework during the back-propagation of errors across trees.

6. Using the Gini impurity measure, compute the impurity reduction ΔI for a split if node t has a class distribution of (p_1, p_2) and the child nodes t_L and t_R have distributions (p_{L1}, p_{L2}) and (p_{R1}, p_{R2}), respectively.

Answers 3

1. **Derive the expression for the change in impurity ΔI:**

$$I(t) = -\sum_{k=1}^{K} p(k \mid t) \log_2 p(k \mid t)$$

Solution: The change in impurity is calculated as:

$$\Delta I = I(t) - \left(\frac{|t_L|}{|t|} I(t_L) + \frac{|t_R|}{|t|} I(t_R) \right)$$

By substituting the entropy expression for each term, we compute:

$$I(t_L) = -\sum_{k=1}^{K} p(k \mid t_L) \log_2 p(k \mid t_L)$$

and similarly for t_R, resulting in an updated calculation for ΔI as:

$$\Delta I = -\sum_{k=1}^{K} \left[p(k \mid t) \log_2 p(k \mid t) - \frac{|t_L|}{|t|} p(k \mid t_L) \log_2 p(k \mid t_L) - \frac{|t_R|}{|t|} p(k \mid t_R) \log_2 p(k \mid t_R) \right]$$

2. **Find the gradient for the loss function** $L(y_i, \hat{y}_i) = (y_i - \hat{y}_i)^2$:
   Solution: The gradient is given by the derivative of the loss function with respect to the predictions $\hat{y}_i$:

$$\frac{\partial L}{\partial \hat{y}_i} = \frac{\partial}{\partial \hat{y}_i}(y_i - \hat{y}_i)^2 = -2(y_i - \hat{y}_i)$$

   Subsequently, the gradient for each data point i is:

$$r_i = -\frac{\partial L}{\partial \hat{y}_i} = 2(y_i - \hat{y}_i)$$

3. **Explain the use of second-order derivatives in gradient boosting**:
   Solution: Second-order derivatives or Hessians provide a measure of the curvature of the loss function:

$$h_i = \frac{\partial^2 L}{\partial \hat{y}_i^2}$$

   In algorithms like XGBoost, the update for parameters uses:

$$\theta \leftarrow \theta - \eta \frac{\sum_i g_i}{\sum_i h_i + \lambda}$$

   where g_i is the first derivative (gradient) and h_i the second derivative (Hessian). This approximation yields more stable and faster convergence by incorporating curvature information, especially beneficial in non-linear contexts.

4. **Derive the expression to initialize the model** $F(x)$:
   Solution: The initial model value γ^* minimizes the empirical risk:

$$\gamma^* = \arg \min_{\gamma} \sum_{i=1}^{n} L(y_i, \gamma)$$

   For squared error loss, this amounts to minimizing:

$$\sum_{i=1}^{n} (y_i - \gamma)^2$$

   Solving $\frac{d}{d\gamma} \sum (y_i - \gamma)^2 = 0$, we find:

$$\gamma^* = \frac{1}{n} \sum_{i=1}^{n} y_i$$

5. **Apply the chain rule in gradient boosting during backpropagation**:
   Solution: In gradient boosting, the differential update for tree predictions necessitates backpropagation, akin to neural networks. Consider a branching decision:

$$\frac{\partial L}{\partial \theta} = \frac{\partial L}{\partial \hat{y}_i} \cdot \frac{\partial \hat{y}_i}{\partial \theta}$$

   For each level:

$$\frac{\partial \hat{y}_i}{\partial \theta} = g(\theta, x_i)$$

   where g represents the tree's structural contribution. Thus, the chain rule accumulates localized tree predictions necessary for back-propagated error updates.

6. **Compute impurity reduction ΔI using Gini impurity:**
   Solution: Gini impurity $I(t)$ is:

$$I(t) = 1 - \sum_k p(k \mid t)^2$$

For a split into t_L and t_R:

$$\Delta I = I(t) - \left(\frac{|t_L|}{|t|} I(t_L) + \frac{|t_R|}{|t|} I(t_R) \right)$$

Substituting:

$$I(t_L) = 1 - \sum_k p_{Lk}^2, \quad I(t_R) = 1 - \sum_k p_{Rk}^2$$

This yields:

$$\Delta I = \left(1 - (p_1^2 + p_2^2)\right) - \left(\frac{|t_L|}{|t|}(1 - (p_{L1}^2 + p_{L2}^2)) + \frac{|t_R|}{|t|}(1 - (p_{R1}^2 + p_{R2}^2)) \right)$$

Chapter 33

Recurrent Neural Networks and Calculus

Practice Problems 1

1. Given the update equation for hidden states in an RNN:

$$h_t = f(W_h h_{t-1} + W_x x_t + b_h)$$

prove that the gradient of the hidden state h_t with respect to the previous hidden state h_{t-1} can be expressed as:

$$\frac{\partial h_t}{\partial h_{t-1}} = f'(W_h h_{t-1} + W_x x_t + b_h)W_h$$

2. Compute the partial derivative of the loss function L with respect to the weight matrix W_h, given:

$$\frac{\partial L}{\partial W_h} = \sum_t \frac{\partial L}{\partial h_t} \frac{\partial h_t}{\partial W_h}$$

Consider the expression for $\frac{\partial h_t}{\partial W_h}$ provided in the chapter.

3. Explain the conditions that lead to vanishing gradients in RNNs and provide a qualitative description of how one could address this issue.

4. Demonstrate how to use the concept of gradient clipping to address the problem of exploding gradients in RNNs.

5. Given the Taylor expansion approximation for the hidden states:

$$h_{t+\Delta} \approx h_t + \Delta t \cdot f'(W_h h_t + W_x x_{t+\Delta} + b_h)$$

verify this approximation using first principles of Taylor series expansion.

6. Discuss the role of calculus in the optimization of learning rates in RNNs using adaptive algorithms like `Adam`, showcasing how the derivatives influence such adaptations.

Answers 1

1. **Solution:** The function h_t is given by:

$$h_t = f(W_h h_{t-1} + W_x x_t + b_h)$$

To find $\frac{\partial h_t}{\partial h_{t-1}}$, we use the chain rule:

The inner function is $u = W_h h_{t-1} + W_x x_t + b_h$ and $h_t = f(u)$.

Hence, the derivative is:

$$\frac{\partial h_t}{\partial h_{t-1}} = \frac{df}{du} \cdot \frac{\partial u}{\partial h_{t-1}} = f'(u) W_h$$

Substituting back, we get:

$$\frac{\partial h_t}{\partial h_{t-1}} = f'(W_h h_{t-1} + W_x x_t + b_h) W_h$$

2. **Solution:** Given the expression for $\frac{\partial L}{\partial W_h}$:

$$\frac{\partial L}{\partial W_h} = \sum_t \frac{\partial L}{\partial h_t} \frac{\partial h_t}{\partial W_h}$$

where:

$$\frac{\partial h_t}{\partial W_h} = \sum_{k \leq t} \frac{\partial h_t}{\partial h_k} \frac{\partial h_k}{\partial W_h}$$

The solution involves summing the contributions from each timestep t, accounting for previous impacts (k) across the temporal sequence.

3. **Solution:** Vanishing gradients occur when gradients shrink rapidly, often when the eigenvalues of W_h are less than one:

$$|\lambda_k(W_h)| < 1$$

One addresses this issue by techniques such as using LSTM units or changing activation functions to those that mitigate shrinkage, preserving gradient flow.

4. **Solution:** Gradient clipping involves limiting the maximum magnitude of gradients to prevent instability in RNNs. Given a gradient vector ∇L,

Clip as follows if $\|\nabla L\| > \theta$:

$$\nabla L \leftarrow \frac{\theta \nabla L}{\|\nabla L\|}$$

Here, θ is a threshold that the magnitude of the gradient should not exceed.

5. **Solution:** The Taylor expansion for functions near a point provides an approximation:

$$h_{t+\Delta} = h_t + \Delta t \cdot u'(t) + \frac{(\Delta t)^2}{2!} \cdot u''(t) + \cdots$$

For a first-order approximation:

$$h_{t+\Delta} \approx h_t + \Delta t \cdot f'(W_h h_t + W_x x_{t+\Delta} + b_h)$$

Confirmation involves showing higher-order terms are negligible for small Δt.

6. **Solution:** Adaptive learning algorithms like `Adam` use first and second moment estimates:

In terms of m_t (mean of gradients) and v_t (variance of gradients):

$$\theta_{t+1} = \theta_t - \alpha \cdot \frac{m_t}{\sqrt{v_t} + \epsilon}$$

Here, derivatives of gradients guide α adjustments, balancing learning rate adaptation and stability.

Practice Problems 2

1. Given the recurrent neural network hidden state equation:

$$h_t = f(W_h h_{t-1} + W_x x_t + b_h)$$

Derive the expression for the gradient $\frac{\partial h_t}{\partial W_h}$.

2. Considering an RNN where the output is defined by:

$$y_t = g(W_y h_t + b_y)$$

Derive $\frac{\partial y_t}{\partial W_y}$.

3. Explain the significance of the vanishing gradient problem in the context of RNNs by calculating $\frac{\partial h_{t-1}}{\partial h_t}$ when using the sigmoid activation function:

$$f(x) = \frac{1}{1 + e^{-x}}$$

4. If the loss function L is defined as:

$$L = \frac{1}{2} \sum_t (y_t - \hat{y}_t)^2$$

Derive $\frac{\partial L}{\partial y_t}$.

5. For a simple RNN with the update rule:

$$h_t = \tanh(W_h h_{t-1})$$

Calculate $\frac{\partial h_t}{\partial W_h}$.

6. Discuss how calculus aids in addressing the exploding gradient problem in RNNs, providing an overview of gradient clipping.

Answers 2

1. For the recurrent neural network hidden state equation:

$$h_t = f(W_h h_{t-1} + W_x x_t + b_h)$$

Solution: Using the chain rule:

$$\frac{\partial h_t}{\partial W_h} = f'(W_h h_{t-1} + W_x x_t + b_h) \cdot h_{t-1}$$

This result arises because the derivative with respect to W_h considers h_{t-1} as constant with respect to weight matrix since W_h acts on the previous hidden state.

2. Given the output equation:

$$y_t = g(W_y h_t + b_y)$$

Solution: The partial derivative is:

$$\frac{\partial y_t}{\partial W_y} = g'(W_y h_t + b_y) \cdot h_t$$

Here, $g'(x)$ represents the derivative of activation function g with respect to its input, emphasizing the direct relationship of each h_t with W_y.

3. For the vanishing gradient problem, using the sigmoid function:

$$f(x) = \frac{1}{1 + e^{-x}}$$

Solution: The derivative of sigmoid is:

$$f'(x) = f(x) \cdot (1 - f(x))$$

Therefore,

$$\frac{\partial h_{t-1}}{\partial h_t} = f'(W_h h_{t-1} + W_x x_t + b_h) \cdot W_h$$

With eigenvalues of W_h less than 1, gradients diminish exponentially over sequences, illustrating the vanishing gradient problem.

4. Given the loss function:

$$L = \frac{1}{2} \sum_t (y_t - \hat{y}_t)^2$$

Solution: The gradient is derived as follows:

$$\frac{\partial L}{\partial y_t} = \frac{\partial}{\partial y_t} \left(\frac{1}{2} (y_t - \hat{y}_t)^2 \right) = (y_t - \hat{y}_t)$$

This shows the basic gradient descent update, driven by the difference between actual and predicted outputs.

5. For the update rule:

$$h_t = \tanh(W_h h_{t-1})$$

Solution: The derivative is:

$$\frac{\partial h_t}{\partial W_h} = \tanh'(W_h h_{t-1}) \cdot h_{t-1}$$

Simplified as:

$$= (1 - \tanh^2(W_h h_{t-1})) \cdot h_{t-1}$$

6. **Solution:** The exploding gradient problem occurs when the gradients become excessively large, destabilizing learning. Calculus addresses this by implementing gradient clipping, where gradients exceeding a threshold are scaled:

$$\text{if } \|\nabla\| > \theta, \quad \nabla = \frac{\theta}{\|\nabla\|} \nabla$$

This ensures stability without losing direction information, allowing effective gradient descent adaptation.

Practice Problems 3

1. Compute the gradient of the hidden state h_t with respect to the weight matrix W_h in the RNN given by:
$$h_t = f(W_h h_{t-1} + W_x x_t + b_h)$$

2. Given the loss function $L = \sum_t \ell(y_t, \hat{y}_t)$, derive the expression for $\frac{\partial L}{\partial W_h}$ in terms of temporal dependencies.

3. Explain how the gradient vanishing problem can be analyzed using the eigenvalues of W_h based on the gradient propagation through time.

4. Derive the update rule in the Backpropagation Through Time (BPTT) algorithm for the RNN's parameters W_h, W_x, and b_h.

5. Explain the purpose of gradient clipping in the context of RNN training and discuss how it alters the gradient descent's convergence behavior.

6. Use a Taylor series to approximate the hidden state $h_{t+\Delta}$ in a time-discrete RNN setting to first order.

Answers 3

1. Compute the gradient of the hidden state h_t with respect to the weight matrix W_h:
   Solution:
   $$\frac{\partial h_t}{\partial W_h} = \frac{\partial}{\partial W_h} f(W_h h_{t-1} + W_x x_t + b_h)$$
   Applying the chain rule:
   $$= f'(W_h h_{t-1} + W_x x_t + b_h) \cdot h_{t-1}$$
   Therefore, the gradient of h_t with respect to W_h is:
   $$\frac{\partial h_t}{\partial W_h} = f'(a_t) \cdot h_{t-1}$$
   where $a_t = W_h h_{t-1} + W_x x_t + b_h$.

2. Given the loss function derive $\frac{\partial L}{\partial W_h}$:
   Solution: Using the chain rule for BPTT:
   $$\frac{\partial L}{\partial W_h} = \sum_t \frac{\partial \ell}{\partial y_t} \frac{\partial y_t}{\partial h_t} \frac{\partial h_t}{\partial W_h}$$
   From the chain of dependencies:
   $$\frac{\partial h_t}{\partial W_h} = \sum_{k \leq t} \frac{\partial h_t}{\partial h_k} \frac{\partial h_k}{\partial W_h}$$
   This gives:
   $$\frac{\partial L}{\partial W_h} = \sum_t \frac{\partial \ell}{\partial y_t} g'(y_t) \sum_{k \leq t} f'(a_k) \cdot h_{k-1}$$

315

3. Explain gradient vanishing using eigenvalues of W_h:
   Solution: The relationship:
   $$\frac{\partial h_t}{\partial h_{t-1}} = f'(a_t)W_h$$

   The vanishing gradient problem arises when:

   $$\prod_{k=1}^{T} |\lambda_k(W_h)| < 1$$

   If all eigenvalues λ_k are small, the product tends to zero, causing gradient magnitudes to vanish over many time steps.

4. Derive the update rule in BPTT:
   Solution: The BPTT update for W_h can be expressed as:
   $$W_h^{new} = W_h - \eta\frac{\partial L}{\partial W_h}$$

   Continuing with the chain rule expansion:

   $$\frac{\partial L}{\partial W_h} = \sum_t \frac{\partial L}{\partial h_t} \sum_{k \leq t} f'(a_t) \cdot h_{t-1}$$

   This yields the update step for BPTT considering all time dependencies.

5. Purpose of gradient clipping:
   Solution: Gradient clipping modifies the parameter update by limiting the maximum norm of the gradient vector:
   $$\nabla L \to \frac{\nabla L}{\|\nabla L\|} \min(\|\nabla L\|, threshold)$$

   Thus preventing $\|\nabla L\| > threshold$ allows for more stable convergence, limiting update size regardless of ∇L.

6. Use Taylor series to approximate $h_{t+\Delta}$:
   Solution: First order Taylor expansion:
   $$h_{t+\Delta} \approx h_t + \Delta t \cdot f'(W_h h_t + W_x x_{t+\Delta} + b_h)$$

   Given small Δt, this provides a linear approximation for progression in time for h.

Made in the USA
Columbia, SC
15 February 2025

53891697R00174